AQA
GCSE chemistry

Author

Philippa Gardom-Hulme

Contents

How to use this book

Welcome to your AQA GCSE Chemistry revision guide. This book has been specially written by experienced teachers and examiners to match the 2011 specification.

On this page you can see the types of feature you will find in this book. Everything in the book is designed to provide you with the support you need to help you prepare for your examinations and achieve your best.

Specification and student book reference: These show how the pages in the unit match to the exam specification and to your textbook so you can track your progress through the unit as you learn.

Key words: These are the terms you need to understand for your exams.

Exam tip: These hints will help you to think about what may come up in the exam.

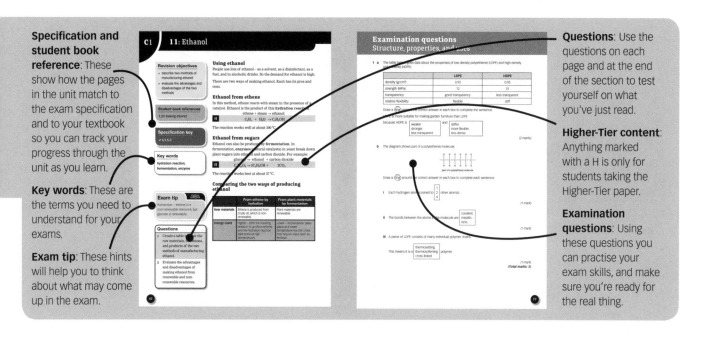

Questions: Use the questions on each page and at the end of the section to test yourself on what you've just read.

Higher-Tier content: Anything marked with a H is only for students taking the Higher-Tier paper.

Examination questions: Using these questions you can practise your exam skills, and make sure you're ready for the real thing.

Upgrade: Upgrade takes you through an exam question in a step-by-step way, showing you why different answers get different grades. Using the tips on this page you can make sure you achieve your best by understanding what each question needs and what an examiner is looking for in your answer.

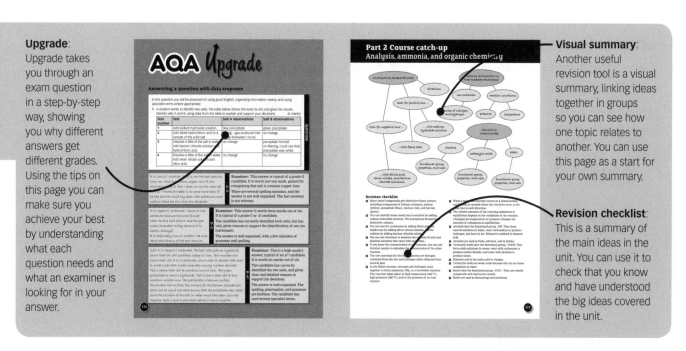

Visual summary: Another useful revision tool is a visual summary, linking ideas together in groups so you can see how one topic relates to another. You can use this page as a start for your own summary.

Revision checklist: This is a summary of the main ideas in the unit. You can use it to check that you know and have understood the big ideas covered in the unit.

Matching your course

The units in this book have been written to match the specification for **AQA GCSE Chemistry**.

In the diagram below you can see that the units and part units can be used to study either for **GCSE Chemistry** or as part of **GCSE Science** and **GCSE Additional Science** courses.

	GCSE Biology	GCSE Chemistry	GCSE Physics
GCSE Science	B1 (Part 1)	C1 (Part 1)	P1 (Part 1)
	B1 (Part 2)	C1 (Part 2)	P1 (Part 2)
GCSE Additional Science	B2 (Part 1)	C2 (Part 1)	P2 (Part 1)
	B2 (Part 2)	C2 (Part 2)	P2 (Part 2)
	B3 (Part 1)	C3 (Part 1)	P3 (Part 1)
	B3 (Part 2)	C3 (Part 2)	P3 (Part 2)

GCSE Chemistry assessment

The units in this book are broken into two parts to match the different types of exam paper on offer. The diagram below shows you what is included in each exam paper. It also shows you how much of your final mark you will be working towards in each paper.

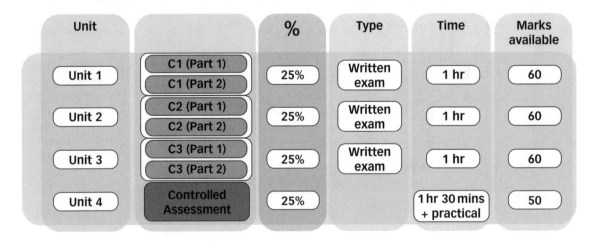

Unit		%	Type	Time	Marks available
Unit 1	C1 (Part 1) / C1 (Part 2)	25%	Written exam	1 hr	60
Unit 2	C2 (Part 1) / C2 (Part 2)	25%	Written exam	1 hr	60
Unit 3	C3 (Part 1) / C3 (Part 2)	25%	Written exam	1 hr	60
Unit 4	Controlled Assessment	25%		1 hr 30 mins + practical	50

When you read the questions in your exam papers you should make sure you know what kind of answer you are being asked for. The list below explains some of the common words you will see used in exam questions. Make sure you know what each word means. Always read the question thoroughly, even if you recognise the word used.

Calculate

Work out your answer by using a calculation. You can use your calculator to help you. You may need to use an equation; check whether one has been provided for you in the paper. The question will say if your working must be shown.

Describe

Write a detailed answer that covers what happens, when it happens, and where it happens. The question will let you know how much of the topic to cover. Talk about facts and characteristics. (Hint: don't confuse with 'Explain')

Explain

You will be asked how or why something happens.Write a detailed answer that covers how and why a thing happens. Talk about mechanisms and reasons. (Hint: don't confuse with 'Describe')

Evaluate

You will be given some facts, data or other information. Write about the data or facts and provide your own conclusion or opinion on them.

Outline

Give only the key facts of the topic. You may need to set out the steps of a procedure or process – make sure you write down the steps in the correct order.

Show

Write down the details, steps or calculations needed to prove an answer that you have been given.

Suggest

Think about what you've learnt in your science lessons and apply it to a new situation or a context. You may not know the answer. Use what you have learnt to suggest sensible answers to the question.

Write down

Give a short answer, without a supporting argument.

Top tips

Always read exam questions carefully, even if you recognise the word used. Look at the information in the question and the number of answer lines to see how much detail the examiner is looking for.

You can use bullet points or a diagram if it helps your answer.

If a number needs units you should include them, unless the units are already given on the answer line.

Revision objectives

- ✓ be able to name the particles that make up atoms
- ✓ link electronic structure to position in the Periodic Table
- ✓ describe some properties of the Group 1 and Group 0 elements

Atoms and elements

All substances are made up of tiny particles called **atoms**. A substance that consists of just one type of atom is an **element**.

There are about 100 elements. They are all listed in the **periodic table**. The vertical columns are called **groups**. The elements in a group have similar properties.

The atoms of each element are represented by a chemical **symbol**. The symbol O represents an atom of oxygen. The symbol Na represents an atom of sodium.

Inside atoms

Atoms are mainly empty space. At the centre of an atom is its **nucleus**. The nucleus is made up of tiny particles called **protons** and **neutrons**. Outside the nucleus are even tinier particles, called **electrons**. The table shows the relative electrical charges of these particles.

Name of particle	Charge
Proton	+1
Neutron	0
Electron	−1

In an atom, the number of protons is the same as the number of electrons. This means that atoms have no overall electrical charge. For example, an atom of the element sodium is made up of:

- 11 protons
- 12 neutrons
- 11 electrons.

The elements on the left of the stepped line are metals.

The elements on the right of the stepped line are non-metals.

The elements in Group 1 are silver-coloured metals. They react with water.

The elements in Group 0 are gases at room temperature. They do not usually react with other substances.

▲ The periodic table.

Atoms of the same element have the same number of protons. Atoms of different elements have different numbers of protons. For example:

- All sodium atoms have 11 protons.
- All oxygen atoms have 8 protons.

The number of protons in an atom of an element is the **atomic number** of an atom. Since sodium has 11 protons, its atomic number is 11.

The sum of the number of protons and neutrons in an atom is its **mass number**. In a sodium atom there are 11 protons and 12 neutrons. This means that the mass number of this sodium atom is 11 + 12 = 23.

Arranging electrons

Electrons are arranged in **energy levels**. Electrons fill the lowest energy levels first. Energy levels are also called **shells**.

Sodium has 11 electrons. Two electrons fill up the lowest energy level. Eight are in the next. The highest energy level contains one electron.

In all atoms, the lowest energy level holds a maximum of two electrons. The next energy level holds up to eight electrons.

Electron arrangements and the periodic table

Elements in the same group of the periodic table have the same number of electrons in their highest energy level. For example, atoms of Group 1 elements have one electron in their highest energy level. Atoms of elements in Group 0 have eight electrons in their highest energy level. Similar electron arrangements give elements similar properties. For example:

- Group 1 elements react vigorously with water to make hydroxides and hydrogen gas:

    potassium + water → potassium hydroxide + hydrogen

    Group 1 elements also react vigorously with oxygen to make oxides:

    potassium + oxygen → potassium oxide

- Group 0 elements are also called the **noble gases**. They are unreactive. They have stable electron arrangements, with eight electrons in the highest energy level, except for helium, which has two.

Key words

atom, element, periodic table, group, symbol, nucleus, proton, neutron, electron, atomic number, mass number, energy level, shell, noble gas

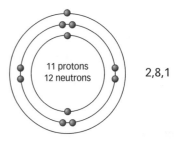

▲ The electronic structure of sodium.

Exam tip

Practise drawing the electron arrangements of the first 20 elements of the periodic table.

Questions

1 List the charges on a proton, a neutron, and an electron.

2 An atom of argon has 18 electrons. Draw its electron arrangement.

3 An atom of fluorine has nine protons, nine electrons, and ten neutrons. Calculate its atomic number and its mass number.

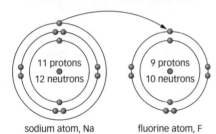

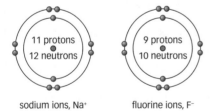

sodium atom, Na — fluorine atom, F

▲ The formation of sodium fluoride.

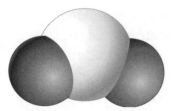

sodium ions, Na⁺ — fluorine ions, F⁻

▲ The electronic structure of sodium and fluoride ions.

▲ A sulfur dioxide molecule.

Compounds

A **compound** is a substance made up of two or more elements. Compounds can be made from their elements, when atoms of the elements join together in chemical reactions.

Holding compounds together

Joining atoms together to form compounds involves giving and taking or sharing electrons.

Ionic bonds

Sodium fluoride is made up of a metal (sodium) joined to a non-metal (fluorine). When sodium fluoride forms from its elements, each sodium atom transfers one of its electrons to a fluorine atom.

Each atom now has eight electrons in its highest energy level. This electron arrangement is very stable.

*   The sodium atom now has 10 electrons and 11 protons. It has a charge of +1.
*   The fluorine atom now has 10 electrons and 9 protons. It has a charge of –1.

Charged atoms are called **ions**. Sodium fluoride contains two types of ion: positive sodium ions (Na^+) and negative fluoride ions (F^-).

Sodium fluoride is an **ionic compound**, like most compounds made up of a metal and a non-metal. In ionic compounds, the oppositely charged ions are strongly attracted to each other. The forces of attraction between the ions are called **ionic bonds**.

Covalent bonds

Compounds formed from non-metals consist of **molecules**. A molecule is a group of atoms held together by shared pairs of electrons. These shared pairs of electrons are **covalent bonds**.

Sulfur dioxide is made up of atoms of two non-metals. Its atoms are joined together in groups of three. Each group consists of one sulfur atom and two oxygen atoms. The atoms are joined together by strong covalent bonds.

Word equations

Word equations summarise chemical reactions. For example, when magnesium burns in oxygen the product is magnesium oxide:

magnesium + oxygen → magnesium oxide

H Symbol equations

Every element and compound has its own formula.
- The formula of nitrogen gas is N_2. This means that nitrogen gas exists as molecules, each made up of two nitrogen atoms.
- The formula of sodium fluoride is NaF. This means that the compound is made up of sodium and fluorine. For every ion of sodium there is one fluoride ion.

When you write a symbol equation for a reaction, start by writing a word equation. Then write down the formula of every element or compound.

magnesium + oxygen → magnesium oxide
$$Mg + O_2 \rightarrow MgO$$

Now balance the equation. There are two atoms of oxygen on the left of the arrow, and one on the right. Write a large '2' to the left of the formula for magnesium oxide:

$$Mg + O_2 \rightarrow 2MgO$$

The number of oxygen atoms is now the same on each side.

Balance the amounts of magnesium by writing a large '2' to the left of the symbol for magnesium. There are now two atoms of magnesium on each side. The equation is balanced.

$$2Mg + O_2 \rightarrow 2MgO$$

Conservation of mass

During a chemical reaction, atoms are not lost or made – so the mass of reactants is equal to the mass of the products. For example, reacting 48 g of magnesium with 32 g of oxygen produces (48 + 32) = 80 g of magnesium oxide.

Exam tip

Practise writing word equations. Practise balancing symbol equations.

Questions

1. Name the type of bonding in nitrogen dioxide and in iron oxide.

2. Write a word equation for the reaction of sodium with chlorine to make sodium chloride.

3. **H** Write a balanced symbol equation for the reaction of lithium with fluorine to form lithium fluoride. Formulae: Li, F_2, LiF.

1   Use the periodic table to give:
    a   the symbol for an atom of oxygen
    b   the name of the element whose atoms have the symbol Fe
    c   the names and symbols of five elements whose names begin with the letter C
    d   the names of four noble gases
    e   the symbols of four elements in Group 1.

2   Name the two types of particle that make up the nucleus of an atom.

3   Complete the sentences below.
    There are about _____ different elements. The _____ in the periodic table contain elements with similar properties.

4   Complete the table to show the electronic structures of some Group 1 elements.

Name of element	Number of electrons	Electronic structure
Lithium	3	
Sodium		2.8.1
Potassium	19	

5   Describe one chemical property that is common to the Group 1 elements.

6   Finish the diagrams below to show the electronic structures of atoms of carbon, silicon, and calcium.

  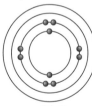

carbon   silicon   calcium
(6 electrons)   (14 electrons)   (20 electrons)

Working to Grade C

7   Neon, argon, and krypton are noble gases.
    a   Give the number of electrons in the outer energy level of an atom of each of these elements.
    b   Name the noble gas that has two electrons in the outer energy level of its atoms.
    c   Explain why the noble gases are unreactive.

8   An atom of an element has five protons and six neutrons.
    a   What is the atomic number of the atom?
    b   What is the mass number of the atom?

9   An atom of an element has an atomic number of 21 and a mass number of 45. How many neutrons are in the nucleus of the atom?

10   Use the periodic table to help you complete the table below.

Name of element	Number of protons	Number of neutrons	Number of electrons
Hydrogen			
Oxygen			
Sodium			
Aluminium			
Boron			
Calcium			

11   Annotate the diagram to represent the electron transfer when sodium reacts with chlorine to form an ionic compound.

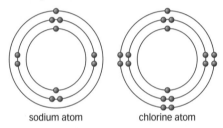

sodium atom   chlorine atom

12   Name the type of bonding in compounds formed from non-metal elements.

13   Compounds that consist of a metal and a non-metal are made of what type of particle?

14   Write word equations to represent the following reactions:
    a   Carbon reacts with oxygen to make carbon dioxide.
    b   Magnesium reacts with chlorine to make magnesium chloride.
    c   Iron reacts with sulfur to make iron sulfide.
    d   Sulfuric acid is neutralised by sodium hydroxide to make sodium sulfate and water.
    e   Lead carbonate decomposes on heating to make lead oxide and carbon dioxide.

15   A chemist burns 3.2 g of sulfur in oxygen gas. She makes 6.4 g of sulfur dioxide, and no other products. What mass of oxygen reacted with the sulfur?

Working to Grade A*

16   State which of the symbol equations below are not balanced. Then balance the equations that are not balanced.
    a $Mg + O_2 \rightarrow MgO$
    b $C + O_2 \rightarrow CO_2$
    c $H_2SO_4 + NaOH \rightarrow Na_2SO_4 + H_2O$
    d $Na_2CO_3 \rightarrow Na_2O + CO_2$
    e $CuSO_4 + Zn \rightarrow Cu + ZnSO_4$
    f $CH_4 + O_2 \rightarrow CO_2 + H_2O$

1 a This diagram shows the electronic structure of an atom.

    i Explain why the element is in Group 6 of the periodic table.

....

(1 mark)

    ii Give the number of protons in the nucleus of the atom above.

....

(1 mark)

b An atom of another element is made up of the particles in the table.

Name of particle	Number
Proton	9
Neutron	10
Electron	9

    i What is the atomic number of the element?

....

(1 mark)

    ii What is the mass number of the element?

....

(1 mark)

    iii Use the periodic table to find out the name and symbol of the element in the table above.

....

(2 marks)

    iv Draw the electronic structure of an atom of the element in the space below.

(2 marks)

(Total marks: 8)

2 This question is about the element lithium and its oxide.

a A teacher heats lithium in oxygen. It reacts to form lithium oxide.

    i Complete the word equation for the reaction.

      lithium + oxygen → ....

                                                                                                     (1 mark)

H   **ii** Balance the symbol equation for the reaction.

        $Li + O_2 \rightarrow Li_2O$

                                                                                      (1 mark)

b Predict the mass of lithium oxide that would be made by reacting 28 g of lithium with 32 g of oxygen.

....

                                                                                               (2 marks)

c Lithium oxide consists of ions.

    i Draw a ring around the correct **bold** word in the sentences below.

      Lithium is a **non metal/metal**.

      A lithium atom **loses/gains** an electron to form an ion with a **negative/positive** charge.

                                                                                                           (3 marks)

    ii The electronic structure of a lithium atom is 2.1.

      Write down the electronic structure of a lithium ion.

....

                                                                                                          (1 mark)

d Use the periodic table to name one element with the same number of electrons in its highest occupied energy level (outermost shell) as a lithium atom.

....

                                                                                                           (1 mark)

e Use the periodic table to write down the number of protons and neutrons in one lithium atom.

    Number of protons = ...

    Number of neutrons = ...

                                                                                                                                             (2 marks)

  (Total marks: 11)

Using limestone

Limestone is a type of rock. It is mainly calcium carbonate, $CaCO_3$.

- Natural limestone is an attractive building material.
- Heating limestone with clay makes **cement**.
  - > Mixing cement with sand makes **mortar**. Builders use mortar to stick bricks together.
  - > Mixing cement with sand and **aggregate** (small stones) makes **concrete**. Concrete is a construction material.

Quarrying

Companies dig limestone from **quarries**. The table shows the impacts of quarrying and using limestone.

	Advantages	Disadvantages
Environmental	Old quarries can be made into lakes.	Quarries make land unavailable for other purposes, such as farming.
Social	Quarries provide jobs and limestone makes buildings and cement.	Quarries create extra traffic.
Economic	Quarry companies sell the limestone.	Tourists may stop visiting an area with a new quarry.

Thermal decomposition

On heating, calcium carbonate breaks down into simpler compounds:

calcium carbonate → calcium oxide + carbon dioxide

H $CaCO_3$ → CaO + CO_2

This is a **thermal decomposition** reaction.

The carbonates of magnesium, zinc, and copper also decompose on heating. The reactions follow a pattern, producing a metal oxide and carbon dioxide.

For example:

copper carbonate → copper oxide + carbon dioxide
(green)   (black)

H $CuCO_3$ → CuO + CO_2

Group 1 metal carbonates do not decompose at Bunsen burner temperatures except for lithium carbonate.

Revision objectives

- identify the benefits and problems of using limestone
- name the products of the thermal decomposition reactions of metal carbonates
- describe the reactions of calcium oxide and calcium hydroxide
- predict the salts formed when carbonates react with acids

Student book references

1.6   Limestone
1.7   The lime cycle
1.8   Products from limestone

Specification key

✓ C1.2.1

Calcium oxide and calcium hydroxide

Adding water to calcium oxide makes calcium hydroxide.

calcium oxide + water → calcium hydroxide

$$\text{H} \qquad CaO \quad + \quad H_2O \quad \rightarrow \qquad Ca(OH)_2$$

Calcium hydroxide dissolves in water. The solution is alkaline. It can be used to neutralise acids.

A solution of calcium hydroxide in water is called **limewater**. Bubbling carbon dioxide gas into colourless limewater makes calcium carbonate. Tiny pieces of solid calcium carbonate make the mixture look cloudy. This is the test for carbon dioxide gas.

calcium hydroxide + carbon dioxide → calcium carbonate + water

$$\text{H} \qquad Ca(OH)_2 + CO_2 \rightarrow CaCO_3 + H_2O$$

Carbonates and acids

Calcium carbonate reacts with sulfuric acid to make calcium sulfate, carbon dioxide, and water. Calcium sulfate is a salt.

calcium carbonate + sulfuric acid → calcium sulfate + carbon dioxide + water

$$\text{H} \qquad CaCO_3 + H_2SO_4 \rightarrow CaSO_4 + CO_2 + H_2O$$

Sulfuric acid is one of the acids in acid rain. Since limestone is mainly calcium carbonate, it is damaged by acid rain.

Other carbonates react with acids in the same way. The products are always carbon dioxide, water, and a salt. Different acids make different salts:

- Hydrochloric acid makes chlorides.
  For example:

  copper carbonate + hydrochloric acid → copper chloride + carbon dioxide + water

  $$\text{H} \quad CuCO_3 + 2HCl \rightarrow CuCl_2 + CO_2 + H_2O$$

- Nitric acid makes nitrates.
  For example:

  zinc carbonate + nitric acid → zinc nitrate + carbon dioxide + water

  $$\text{H} \qquad ZnCO_3 + 2HNO_3 \rightarrow Zn(NO_3)_2 + CO_2 + H_2O$$

- Sulfuric acid makes sulphates.
  For example:

  magnesium carbonate + sulfuric acid → magnesium sulfate + carbon dioxide + water

  $$\text{H} \qquad MgCO_3 + H_2SO_4 \rightarrow MgSO_4 + CO_2 + H_2O$$

Exam tip

Practise writing word equations for all the reactions in this topic.

Questions

1 Name the substances that are mixed to make mortar and concrete.

2 Identify the products of the reaction of calcium hydroxide solution with carbon dioxide gas.

3 Write word equations for the reactions of:

   a calcium oxide with water

   b magnesium carbonate with nitric acid.

1 Limestone is quarried from the ground. Suggest:
   a a social advantage of quarrying limestone
   b a social disadvantage of quarrying limestone
   c an environmental advantage of quarrying limestone
   d an economic disadvantage of quarrying limestone.

2 Complete the sentences below.
   Cement is made by heating together two raw materials: limestone and _____. Mixing cement with sand makes _____. Mixing cement with sand and aggregate makes _____.

3 Draw a ring around the one carbonate in the list below that cannot be decomposed at the temperatures reached by a Bunsen burner:

   potassium carbonate

   magnesium carbonate

   copper carbonate

4 Describe how to use a solution of calcium hydroxide in water (limewater) to test for carbon dioxide gas. Describe all the changes you would expect to see.

5 Complete the flow diagram below.

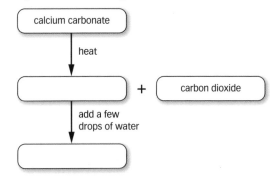

6 Calcium carbonate decomposes on heating to make calcium oxide and carbon dioxide. Write a word equation for the reaction.

7 Write word equations for the thermal decomposition reactions of:
   a zinc carbonate
   b lead carbonate.

8 Name the salts formed when the following pairs of substances react together:
   a copper carbonate and dilute sulfuric acid
   b zinc carbonate and dilute hydrochloric acid
   c calcium carbonate and dilute nitric acid
   d magnesium carbonate and dilute hydrochloric acid
   e sodium carbonate and dilute sulfuric acid.

9 Complete the following word equations:
   a sodium carbonate + nitric acid →
   b magnesium carbonate + nitric acid →
   c calcium carbonate + hydrochloric acid →
   d zinc carbonate + sulfuric acid →

10 A student collected secondary data about the properties of four building materials. The data is in the table below. The student also made notes about their properties. The notes are below the table.

Material	Tensile strength (MPa)	Compressive strength (MPa)	Thermal conductivity (W/m/k)
Mild steel	250	250	60.00
Oak wood	21	15	0.16
High strength concrete	Cracks if not reinforced	60	1.45
Limestone	No data	102	1.30

Notes:
- Tensile strength measures the force needed to pull something to the point where it starts breaking.
- Compressive strength measures the pushing force needed to crush a material.
- Thermal conductivity describes how well a material conducts heat. The higher the number, the better it conducts heat.

   a Which building material can withstand the greatest force before being crushed?
   b Which two materials conduct heat least well? Suggest why this might be a desirable property for a building material.
   c Explain why the student cannot be sure which of the four materials has the greatest tensile strength.

11 Complete the equation below for the thermal decomposition reaction of zinc carbonate. The formulae of the products are ZnO and CO_2.
   $$ZnCO_3 \rightarrow$$

12 Bubbling carbon dioxide gas through a dilute solution of calcium hydroxide produces calcium carbonate and water.
   Complete the equation below to summarise the reaction.
   $$Ca(OH)_2 + CO_2 \rightarrow$$

13 Check the equations below to see if they are balanced.
   Balance those that are not already balanced.
   a $MgCO_3 + HNO_3 \rightarrow Mg(NO_3)_2 + CO_2 + H_2O$
   b $ZnCO_3 + HCl \rightarrow ZnCl_2 + CO_2 + H_2O$
   c $Na_2CO_3 + HCl \rightarrow NaCl + CO_2 + H_2O$
   d $CuCO_3 + H_2SO_4 \rightarrow CuSO_4 + CO_2 + H_2O$
   e $CaCO_3 + HNO_3 \rightarrow Ca(NO_3)_2 + CO_2 + H_2O$

1 a Every year, UK companies extract millions of tonnes of limestone from quarries.

    i Suggest an economic advantage of quarrying limestone.

....

(1 mark)

    ii Suggest an environmental disadvantage of quarrying limestone.

....

(1 mark)

b Limestone is mainly calcium carbonate.

When heated in lime kilns, calcium carbonate decomposes to make calcium oxide (a useful product) and carbon dioxide (a waste gas).

    i Write a word equation for the thermal decomposition reaction of calcium carbonate.

....

(1 mark)

    ii Suggest why traditional lime kilns were open to the air.

....

(1 mark)

    iii Calcium oxide reacts with water to make calcium hydroxide.

       Give two uses of calcium hydroxide, either as a solid or in solution.

       1 ....

       2 ....

(2 marks)

c This statue is made of dolomite rock.

Dolomite rock is mainly calcium magnesium carbonate.

Predict the names of two salts formed when acid rain falls on the statue. Acid rain is a mixture of dilute acids, including sulfuric acid.

....

....

(2 marks)

H

d Balance the equation below.

It shows the reaction of calcium carbonate with hydrochloric acid.

$CaCO_3 + HCl \rightarrow CaCl_2 + CO_2 + H_2O$

(2 marks)

(Total marks: 10)

2 A student investigated the thermal decomposition reactions of some metal carbonates. He heated each metal carbonate in turn in the apparatus below.

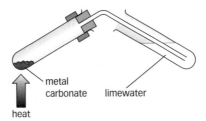

His results are in the table below.

Name of metal carbonate	Time for limewater to begin to look cloudy (seconds)
Sodium carbonate	Did not go cloudy
Calcium carbonate	275
Magnesium carbonate	153
Zinc carbonate	50
Copper carbonate	20

a i Identify **two** control variables for the investigation.

1 ....

2 ....

(2 marks)

ii Name the dependent variable in the investigation.

....

(1 mark)

b On heating, zinc carbonate decomposed to make zinc oxide and carbon dioxide gas. Write a word equation for this reaction.

....

(1 mark)

c Name the two products of the thermal decomposition reaction of magnesium carbonate.

.... and ....

(2 marks)

d Complete the word equation for the reaction that makes the limewater go cloudy. Limewater is a solution of calcium hydroxide in water.

calcium hydroxide + ... → .... + water

(2 marks)

(Total marks: 7)

Revision objectives

- ✓ outline how iron and aluminium are extracted from their ores
- ✓ describe new techniques of copper extraction

Student book references

1.9 Magnificent metals

1.10 Stunning steel

1.11 Copper

Specification key

✓ C1.3.1 a–h

sodium
calcium
magnesium
aluminium
carbon
zinc
iron
tin
lead
copper
silver
gold

▲ The reactivity series.

Metal ores

An **ore** is a rock from which a metal can be extracted economically. When deciding whether to extract a metal from a particular ore, companies consider:

- the price they can get for the metal
- the costs of extracting the metal.

Economic factors change over time. If the demand for a metal increases, companies may make a profit by extracting it from an ore that contains a smaller proportion of the metal.

Processing ores

Companies dig metal ores from the ground. This is **mining**. The ore is then **concentrated** to separate its metal compounds from waste rock. Next, the metal is extracted from its compounds, and purified.

Extracting metals

How a metal is extracted depends on its position in the reactivity series.

- **Unreactive** metals, such as gold, exist in the Earth as the metals themselves, not as compounds.
- Most metals exist as compounds:
  - > Metals that are less reactive than carbon, such as iron, are extracted from their oxides by **reduction** with carbon.
  - > More reactive metals, such as aluminium, are extracted by the **electrolysis** of their molten compounds.

Extracting iron

Iron is extracted from its oxides in a **blast furnace**. Here's how:

- Put the iron ore (mainly iron(III) oxide, Fe_2O_3) into a hot blast furnace with carbon.
- Reduction reactions remove oxygen from the iron(III) oxide. The products are iron and carbon dioxide.

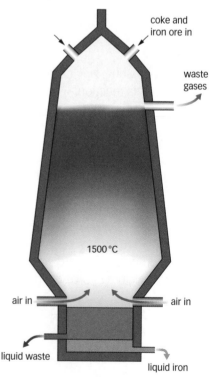

coke and iron ore in

waste gases

1500 °C

air in

air in

liquid waste

liquid iron

▲ Iron oxide is reduced in the blast furnace to make iron.

Extracting aluminium

Aluminium is extracted by electrolysis. Electricity passes through liquid aluminium oxide. The aluminium oxide breaks down and:

- positive aluminium ions move to the negative electrode; aluminium metal is produced here.
- negative oxide ions move to the positive electrode.

The process needs large amounts of energy, so aluminium is expensive.

Extracting copper from copper-rich ores

Companies heat copper-rich ores in a furnace. Chemical reactions remove other elements from the copper compounds in the ore. This is **smelting**.

The copper is then purified by **electrolysis**. The diagram on the right shows how.

Extracting copper from low-grade ores

Copper-rich ores are running out. So companies now extract copper by smelting **low-grade ores** that contain less copper. This is expensive, but the high demand for copper means companies can still make a profit. Extracting copper from low-grade ores produces huge amounts of waste. This damages the environment.

Phytomining and bioleaching

Scientists have developed new ways of extracting copper from low-grade ores. The techniques damage the environment less. They include:

- **Phytomining** – This involves planting certain plants on low-grade copper ores. The plants absorb copper compounds. Burning the plants produces an ash that is rich in copper compounds.
- **Bioleaching** – This uses bacteria to produce solutions of copper compounds. Copper metal is extracted from these solutions by chemical reactions or electrolysis.

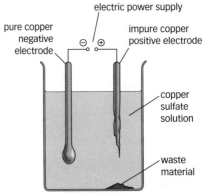

▲ The negative electrode is made of pure copper. Impure copper forms the positive electrode. During electrolysis, positive copper ions move to the negative electrode. Waste material falls to the bottom.

Questions

1. List the four main stages involved in obtaining iron from its ore.

2. Explain why aluminium is extracted from its ore by electrolysis, and not by reduction with carbon.

3. Describe two techniques of extracting copper from low-grade ores.

Exam tip

Remember:
- Metals that are less reactive than carbon are extracted from their oxides by reduction with carbon.
- Metals that are more reactive are extracted by electrolysis.

Revision objectives

- explain the benefits of recycling metals, and describe how copper is recycled
- link the properties of metals and alloys to their uses

Student book references

1.11 Copper

1.12 Titanium and aluminium

Specification key

- ✔ C1.3.1 i–j, ✔ C1.3.2,
- ✔ C1.3.3

Reasons for recycling

Recycling scrap metal involves melting metals and making them into new things. Governments encourage recycling because:

- some metal ores are in short supply
- extracting metals from ores creates waste and requires much energy, so is damaging to the environment.

Titanium metal cannot be extracted from its ore by reduction with carbon. The extraction of titanium is expensive because it has many stages and requires much energy. It makes economic sense to recycle titanium.

Recycling copper

One method of obtaining pure copper from scrap copper involves making solutions of copper salts. Copper metal is extracted from these solutions by:

- **Electrolysis** – positive copper ions move towards the negative electrode. Here, they gain electrons to form copper atoms.
- **Displacement** – scrap iron is added to the solution. Iron is more reactive than copper, so it displaces copper from its compounds. For example:

    copper sulfate + iron → copper + iron sulfate

Using iron

Iron from the blast furnace contains about 96% iron and 4% impurities. The impurities make it brittle – it breaks if you drop it – so it has few uses.

Some blast furnace iron is re-melted and mixed with scrap steel. This makes **cast iron**. Cast iron has a high strength in compression – you can press down on it with a great force and it will not break. Cast iron is used to make cooking pots, and has been used to make arched bridges.

Steels – vital alloys

Steel is mainly iron. The iron is mixed with carbon and other metals to change its properties and make it more useful. There are many types of steel. Each has properties that make it suitable for different uses.

Steels are examples of **alloys**. An alloy is a mixture of a metal with one or more other elements. The properties of an alloy are different to those of the elements in it.

Type of steel	Property	Uses
Low-carbon steel	Easily shaped	Food cans, Car body panels
High-carbon steel	Hard	Tools
Stainless steel	Resistant to corrosion (does not rust)	Cutlery, Surgical instruments

▲ A model of the structure of steel. The bigger circles represent iron atoms. The smaller ones represent atoms of another element.

Other alloys

Metals like copper, gold, iron, and aluminium are too soft for many uses. So they are mixed with small amounts of similar metals to make alloys. The alloys are harder and more useful.

Properties and uses of metals

The elements in the central block of the periodic table are the **transition metals**.

																		0
1	2					H						3	4	5	6	7		He
Li	Be											B	C	N	O	F		Ne
Na	Mg											Al	Si	P	S	Cl		Ar
K	Ca	Sc	Ti	V	Cr	Mn	Fe	Co	Ni	Cu	Zn	Ga	Ge	As	Se	Br		Kr
Rb	Sr	Y	Zr	Nb	Mo	Tc	Ru	Rh	Pd	Ag	Cd	In	Sn	Sb	Te	I		Xe
Cs	Ba	La	Hf	Ta	W	Re	Os	Ir	Pt	Au	Hg	Tl	Pb	Bi	Po	At		Rn
Fr	Ra	Ac																

transition metals

Like most metals, the transition metals:
- can be bent or hammered into different shapes without cracking
- are good conductors of heat and electricity.

These properties make transition metals useful as structural materials, and for making things that must allow heat or electricity to pass through them.

For example, copper is used for:
- electrical wiring, because it is a good conductor of electricity
- plumbing, because it can be bent, but is hard enough to make pipes and tanks. Also, copper does not react with water.

Titanium and aluminium are useful metals. They have low densities and are resistant to corrosion. These properties mean that both metals are used to make aeroplanes. Aluminium makes overhead electricity cables. Titanium makes artificial hip bones.

Questions

1   Give three reasons for recycling metals.

2   Name three types of steel. Link their properties to their uses.

3   Explain why titanium is a suitable metal from which to make artificial hip bones and aeroplanes.

1 Name one metal that is found in the Earth as the metal itself.

2 The list below gives four stages of producing a pure metal from its ore. Put them in the correct order.
   a Purify the metal.
   b Mine the ore.
   c Concentrate ore.
   d Extract the metal from the ore.

3 Use the periodic table to identify the transition metals in the list below.

   molybdenum   sodium
   manganese   aluminium
   magnesium   titanium
   scandium

4 Give two benefits of producing aluminium drinks cans from recycled aluminium, compared to using aluminium that has recently been extracted from its ore.

5 a Complete the table below to show the properties of three types of steel.

Type of steel	Properties
Stainless steel	
High-carbon steel	
Low-carbon steel	

   b List one use of each type of steel in the table.

6 Highlight the correct bold word in the sentences below.
   Iron from the blast furnace is **brittle/ shatterproof**. Re-melting blast furnace iron and adding scrap steel makes **steel/cast iron**. This has high strength in **compression/tension**, so is used to make arched bridges.

7 Explain why titanium and its alloys are used to make:
   a aeroplanes
   b the supports for North Sea oil rigs.

8 In the blast furnace, iron oxide is reduced by heating with carbon. Explain the meaning of the word **reduced**.

9 The metals below are listed in order of their reactivity, with the most reactive at the top. One non-metal, carbon, is also included in the list.

   sodium   iron
   calcium   tin
   magnesium   lead
   aluminium   gold
   carbon

   a Predict two metals, other than iron, that can be extracted from their oxides by reduction with carbon.
   b Predict two metals that are extracted by the electrolysis of their molten compounds.
   c Identify one element that is found in the Earth as the element itself.

10 Explain why it is expensive to extract titanium metal from its ore.

11 Identify two advantages of extracting copper from low-grade ores by phytomining and bioleaching, compared with extracting copper from low-grade ores by smelting.

12 The diagram below shows how copper is purified by electrolysis.

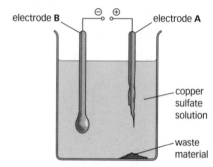

   a Which electrode is impure copper?
   b Towards which electrode do positively charged copper ions move?

13 Write a word equation to summarise how copper metal is extracted from copper sulfate solution by adding scrap iron.

14 The table below gives some data for aluminium and two of its alloys.

Material	Tensile strength (MPa)	Hardness (Brinell scale)*	Density (g/cm³)
Pure aluminium	90	23	2.7
Aluminium alloy 7075	572	150	2.8
Aluminium alloy 5059	160	120	2.7

*The bigger the number, the harder the material.

   a Name the hardest material in the table.
   b Aluminium alloy 7075 is used in making aeroplanes. Use the data in the table to suggest two reasons for this choice.
   c Suggest one disadvantage of using aluminium alloy 7075 for making aeroplanes, compared to using pure aluminium.

Examination questions
Metals and their uses

1 This question is about copper.

a Describe two properties of copper that make it a suitable material for water pipes.

1 ....

2 ....

(2 marks)

b This table shows some data about three materials.

Name of material	Composition
Copper	100% copper
Nickel brass	70% copper, 24.5% zinc, 5.5% nickel
Naval brass	60% copper, 39.25% zinc, 0.75% tin

i Name the alloys in the table.

....

(1 mark)

ii Pound coins are made from nickel brass.

Suggest one advantage of making coins from nickel brass, compared to pure copper.

....

(1 mark)

c **i** Copper can be extracted by mining low-grade ores and heating the ores in a furnace.

Describe two disadvantages of obtaining copper by this method.

1 ....

2 ....

(2 marks)

ii Describe how copper can be extracted by phytomining.

....

....

(2 marks)

(Total marks: 8)

2 Read the article in the box below.

> Large deposits of a rare metal, indium, have been discovered in the UK in an old tin mine in Cornwall. The mine owners hope to dig out millions of pounds worth of indium every year.
>
> Indium tin oxide is used in touch-screen technology and to make liquid-crystal displays for flat-screen televisions, smart phones, and laptops. As more and more people have bought these products over the past 10 years, so the amount of indium needed has increased.

a Scientists analysed samples of rock from the mine. The table below shows data from six samples.

Sample number	Mass of indium per tonne of rock (g)
1	1 000
2	110
3	100
4	90
5	95
6	105

i The scientists concluded that the average mass of indium per tonne of rock is 100 g.

Explain how the data in the table support this conclusion.

....

(1 mark)

ii The mine owner hopes that 400 000 tonnes of rock will be mined each year.

Estimate the mass of indium in this mass of rock.

Assume that the average mass of indium per tonne of rock is 100 g. Give your answer in tonnes. (1 tonne = 1 000 000 g)

....

....

(2 marks)

iii Use your answer to part **ii** to estimate the mass of solid waste produced if 400 000 tonnes of rock are mined each year.

....

....

(1 mark)

iv Suggest an environmental impact of the waste rock.

....

(1 mark)

b The local council must decide whether to allow the mine to reopen to extract indium. People are discussing their ideas.

Extracting indium from the mine could create 400 jobs.

Barney – local resident

China is the biggest producer of indium. In 2009, a quarter of all indium production in the country was stopped. People were protesting about pollution linked to the process.

Lim – Chinese journalist

Our company uses copper indium gallium selenide to make solar cells. As the demand for solar cells increases, so will the demand for indium. The world needs to mine all the indium that is found.

Sarah – solar-cell manufacturer

Indium metal is harmful if swallowed or breathed in.

Meera – British scientist

Last year, 480 tonnes of indium was produced from mining, and 650 tonnes from recycling. We need to recycle more indium, so as not to put the health of humans and the environment at risk.

Catherine – environment worker

Use all the evidence given in this question to recommend whether or not the mine should be allowed to re-open to extract indium.

....

....

....

....

....

(4 marks)

(Total marks: 9)

Crude oil

Crude oil is a **fossil fuel**. It was formed from the decay of buried dead sea creatures over millions of years.

Crude oil contains many different compounds. Many of the compounds are **hydrocarbons**. Hydrocarbons are compounds made up of carbon and hydrogen only.

The hydrocarbons in crude oil form a **mixture**. A mixture consists of two or more elements or compounds that are not chemically combined together. In a mixture:

- the chemical properties of each substance are not affected by being in the mixture – their properties are unchanged
- the substances can be separated by physical methods, including filtering and distillation.

Fractional distillation

Crude oil is not useful as it is. But separating it into **fractions** makes valuable fuels and raw materials. A fraction is a mixture of hydrocarbons whose molecules have a similar number of carbon atoms.

Oil companies use the property of boiling point to separate crude oil into fractions by **fractional distillation**. The process is continuous – it carries on all the time.

Key words

hydrocarbon, mixture, fraction, fractional distillation, evaporate, fractionating column, condense, alkanes, saturated hydrocarbon, molecular formula, displayed formula, general formula

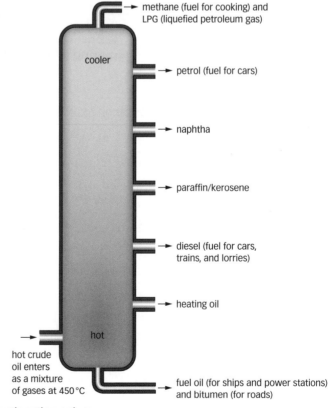

▲ Fractionating column.

Fractional distillation involves these steps:
- Heat crude oil to about 450 °C. Its compounds **evaporate** to become a mixture of gases.
- The gases enter the bottom of a **fractionating column**. The fractionating column is hot at the bottom and cooler at the top.
- The gases move up the column. As they move up, they cool down. Different fractions **condense** to form liquids again at different levels:
  - > Compounds with the highest boiling points condense at the bottom of the column, and leave as liquids.
  - > Lower-boiling-point compounds condense higher up, where it is cooler, and leave as liquids.
  - > The lowest-boiling-point compounds leave from the top, as gases.

Alkanes

Most of the hydrocarbons in crude oil are **alkanes**. Alkanes are a family of **saturated hydrocarbons**. The carbon atoms in saturated hydrocarbons are joined together by single covalent bonds.

Alkane molecules can be represented by **molecular formulae** and by **displayed formulae**. The table below gives the names and formulae of three alkanes. There are many others.

Name	Molecular formula	Displayed formula
Methane	CH_4	H—C—H (with H above and below)
Ethane	C_2H_6	H—C—C—H (with H above and below each C)
Propane	C_3H_8	H—C—C—C—H (with H above and below each C)

In displayed formulae, each line (–) represents a covalent bond.

All alkanes have the same **general formula**, C_nH_{2n+2}. This means that the number of hydrogen atoms in an alkane molecule is twice the number of carbon atoms, plus two.

Exam tip

Make sure you can recognise alkanes from their names and formulae. Practise writing both types of formulae for methane, ethane, and propane.

Questions

1 Write definitions for these words: hydrocarbon, mixture, fraction, alkane.

2 What is a mixture? Give two characteristics of mixtures.

3 Write the molecular formulae of methane, ethane, and propane.

Revision objectives

✓ link the properties of hydrocarbons to the size of their molecules

✓ describe the impacts on the environment of burning hydrocarbon fuels

Student book references

1.14 Looking into oil

1.15 Burning dilemmas

Specification key

✓ C1.4.2 c, ✓ C1.4.3 a–d

Sizes and properties of hydrocarbons

Hydrocarbon properties depend on their molecule size. These properties influence how hydrocarbons are used as fuels.

- Hydrocarbons with bigger molecules are more **viscous** (thicker) than hydrocarbons with smaller molecules. Viscous liquids are more difficult to pour and do not flow as easily as runnier liquids.
- Hydrocarbons with bigger molecules have higher boiling points than those with smaller molecules.

Number of carbon atoms in hydrocarbon chain	State at room temperature
1 to 4 (smaller molecules)	gas
5 to 16	liquid
17 or more (longer molecules)	solid

Gases make good fuels for cooking because they travel easily through pipes. Liquid fuels are easy to store and transport, so are suitable for use as vehicle fuels.

- Alkanes with smaller molecules catch fire more easily than those with bigger molecules. Methane has small molecules, so it ignites easily and is useful for cooking.

▲ Longer-chain alkanes are more viscous because their molecules get more tangled.

Burning hydrocarbon fuels

The **combustion** (burning) of fuels releases energy. During combustion, carbon and hydrogen atoms from fuels join with oxygen from the air. The carbon and hydrogen are **oxidised**.

Burning fuels release gases to the air.

- In a good supply of air, a burning hydrocarbon produces two main products – carbon dioxide and water vapour. For example:

  methane + oxygen → carbon dioxide + water

  H $CH_4 + 2O_2 \rightarrow CO_2 + 2H_2O$

  This is **complete combustion**.

- Burning hydrocarbons in a poor supply of air makes **carbon monoxide** gas (CO) as well as carbon dioxide and water vapour. This is **partial** or **incomplete combustion**.

Carbon monoxide and carbon dioxide cause problems:

- Carbon monoxide is poisonous.
- Carbon dioxide is a **greenhouse gas**. Its presence in the atmosphere helps to keep the Earth warm enough for life. The diagram on the next page shows how.

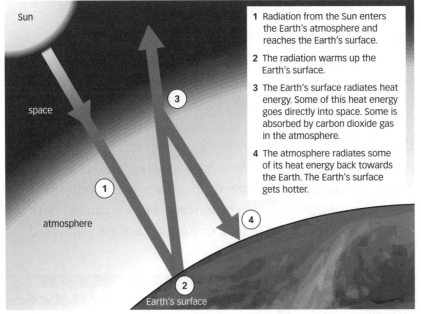

Sun

space

atmosphere

Earth's surface

1 Radiation from the Sun enters the Earth's atmosphere and reaches the Earth's surface.

2 The radiation warms up the Earth's surface.

3 The Earth's surface radiates heat energy. Some of this heat energy goes directly into space. Some is absorbed by carbon dioxide gas in the atmosphere.

4 The atmosphere radiates some of its heat energy back towards the Earth. The Earth's surface gets hotter.

▲ Global warming.

- Extra carbon dioxide in the atmosphere from burning fuels causes **global warming**. The impacts of global warming include:
  - > **climate change**, causing extreme weather events
  - > the melting of the polar ice caps, causing sea levels to rise and coastal areas to flood.

More products of combustion

Some fuels contain atoms of sulfur, as well as carbon and hydrogen. These fuels make **sulfur dioxide** (SO_2) when they burn. At the high temperatures reached in car engines, **oxides of nitrogen**, such as nitrogen monoxide (NO) and nitrogen dioxide (NO_2), also form.

Sulfur dioxide and oxides of nitrogen cause **acid rain**. Acid rain:

- makes lakes more acidic, killing water plants and animals
- damages trees
- damages limestone buildings by reacting with calcium carbonate.

Sulfur can be removed from fuels before they are burnt. Sulfur dioxide can be removed from the waste gases of power stations.

Particulates

Burning fuels may release solid particles, or **particulates**. Particulates contain soot (a form of carbon) and unburnt fuels. Particulates cause **global dimming**. In the atmosphere, particulates reflect sunlight back into space, meaning that less sunlight reaches the Earth's surface.

Exam tip

Make sure you know which environmental problems are caused by which products of combustion – it's easy to get them all confused!

Questions

1 Describe how the boiling points, viscosity, and flammability of hydrocarbons are linked to molecule size.

2 Draw a table to summarise how burning fuels produce these substances: carbon dioxide, carbon monoxide, sulfur dioxide, oxides of nitrogen, particulates. Identify the environmental problems caused by each substance.

Revision objectives

✓ identify the benefits, drawbacks, and risks of using biofuels and hydrogen fuel

Student book references

1.16 Global warming
1.17 Biofuels

Specification key

✓ C1.4.3 e

Finite resources

Crude oil, coal, and natural gas are **non-renewable**, since they formed over millions of years. Fossil fuels are **finite** resources – they will not last for ever.

Biofuels

Biofuels are fuels made from substances obtained from living things. **Ethanol** can be made from sugar cane. **Biodiesel** is made from plant oils such as sunflower oil or palm oil. Both fuels are liquid at normal temperatures. They can be stored and transported safely.

	Advantages	Disadvantages
Economic issues	Biofuel producers sell their products.	Converting filling stations to dispense biofuels is expensive.
Social and ethical issues	Biofuels are renewable, so using them does not take supplies from future generations.	Fuel crops may be grown on land that some people think should be used to grow food.
Environmental issues	The plants from which biofuels are made remove carbon dioxide from the atmosphere as they grow.	Overall, more carbon dioxide is added to the atmosphere than is removed, since carbon dioxide is released by farm machinery and during the production of fertilisers.

Hydrogen fuel

Hydrogen can also be used as a vehicle fuel. It ignites easily, and produces just one product on burning – water vapour. Hydrogen is explosive when mixed with air, so is difficult to store and transport safely.

Hydrogen can be made from methane gas, which is renewable when produced from animal waste.

Key words

non-renewable, finite, biofuel, ethanol, biodiesel

Exam tip

Read questions on this topic carefully, and answer in detail.

Questions

1 Evaluate the environmental impacts of using biodiesel as a vehicle fuel instead of diesel.

2 List the benefits and drawbacks of using hydrogen gas as a vehicle fuel.

1 Highlight the statements below that are true. Then write corrected versions of the statements that are false.

   a Crude oil is a mixture of elements.

   b A mixture consists of two or more elements or compounds that are chemically combined.

   c When substances are mixed together, their chemical properties change.

   d You can separate the substances in a mixture by physical methods, such as distillation.

   e Crude oil is separated into fractions by fractional distillation.

2 In each sentence below, highlight the correct **bold** word.

   a Most compounds in crude oil are **hydrocarbons/carbohydrates**.

   b Most compounds in crude oil are made up of carbon and **hydrogen/helium** only.

   c Most compounds in crude oil are **alkenes/alkanes**.

   d Alkanes are **saturated/unsaturated** compounds.

3 The statements below describe some stages in the fractional distillation of crude oil. Write the letters of the stages in the correct order.

   a The compounds in the oil evaporate.

   b The mixture of vapours enters the fractionating column.

   c The mixture of vapours moves up the fractionating column.

   d The oil is heated.

   e Fractions of hydrocarbons with the highest boiling points condense near the bottom of the fractionating column.

   f Fractions of hydrocarbons with the lowest boiling points are removed near the top of the fractionating column.

4 Draw lines to match the name of each pollutant to show how it is formed.

Pollutant	Formed when ...
Nitrogen dioxide	hydrocarbons burn in a poor supply of air
Carbon monoxide	hydrocarbons burn in air at high temperatures
Sulfur dioxide	a sulfur-containing fuel burns

5 Name one environmental problem caused by each of these substances:

   a carbon dioxide

   b nitrogen monoxide

   c sulfur dioxide

   d particulates (solid particles)

6 Complete the table below.

Name of alkane	Molecular formula	Structural formula
Methane		
	C_2H_6	
	C_4H_{10}	

7 Use either the word **increases** or **decreases** to complete each sentence below.

   a The boiling point of alkanes increases as their molecule size ....

   b Alkanes get less viscous as their molecule size ....

   c Alkanes get more flammable as their molecule size ....

8 Complete these word equations:

   a The complete combustion of butane.
   butane + oxygen →

   b The partial combustion of methane.
   methane + ... → carbon dioxide + ... + ...

9 The table below shows the temperature change of 100 cm³ of water when heated by burning the same mass of four different alkanes.

Alkane	Temp before heating (°C)	Temp after heating (°C)	Temp change (°C)
Methane	19	75	56
Ethane	20	72	
Propane	20	68	48
Butane	21	71	50

   a Calculate the temperature rise of the water when heated by ethane.

   b Which fuel transferred most energy to the water?

   c What happens to the atoms of carbon and hydrogen when alkanes burn?

10 Complete the table below to show the benefits and drawbacks of using ethanol as a fuel, compared to using petrol. Assume the ethanol is produced from plant material and the petrol is produced by the fractional distillation of crude oil.

	Ethanol	Petrol
Use of renewable resources	Pro: Con:	Pro: Con:
Fuel storage and use	Pro: Con:	Pro: Con:
Combustion products	Pro: Con:	Pro: Con:

1 This question is about the substances released into the atmosphere when fuels burn.

    a Draw lines to link each substance to the problem it may cause. One problem is caused by more than one substance.

Substance
carbon dioxide
sulfur dioxide
solid particles
oxides of nitrogen

Problem
acid rain
global warming
global dimming

(4 marks)

    b **i** Write a word equation for the complete combustion of methane to form carbon dioxide and water.

....

(2 marks)

        ii Name the carbon compound formed as a result of the partial combustion of methane.

....

(1 mark)

(Total marks: 7)

2 The compounds in crude oil are separated by fractional distillation. This diagram shows a fractionating column.

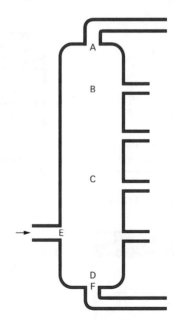

a Use words from the box to complete the following sentences. Each word may be used once, more than once, or not at all.

| cool | evaporate | condense | freeze | warm |

The substances in crude oil .... to become a mixture of gases. Fractions in the mixture

.... at different temperatures as they ....

(3 marks)

b Match the labels in the table below with the correct letter in the diagram of the fractionating column. Write the correct letter beside each label. You can use each letter once, more than once, or not at all.

	Letter
The place where a mixture of vapours enters the column.	
The hottest part of the column.	
The place where the fraction containing substances with the highest boiling points leaves the column.	
The place where the fraction containing the most flammable substances leaves the column.	
The place where methane gas leaves the column.	

(5 marks)
(Total marks: 8)

3 This table gives data for six alkanes.

Name of alkane	Molecular formula	Melting point (°C)	Boiling point (°C)
Methane	CH_4	−182	−162
Ethane	C_2H_6	−183	−89
Propane	C_3H_8	−188	−42
Butane	C_4H_{10}	−138	−0.5
Pentane	C_5H_{12}	−130	36
Hexane	C_6H_{14}	−95	69

a Draw the displayed formula for propane in the space below.

(2 marks)

b i Describe the trend in boiling points shown in the table.

....

(1 mark)

ii Name an alkane in the table that is a liquid at room temperature (20 °C).

....

(1 mark)
(Total marks: 4)

4 *In this question you will be assessed on using good English, organising information clearly, and using specialist terms where appropriate.*

Describe the benefits, drawbacks, and risks of using hydrogen as a vehicle fuel, compared with using hydrocarbon fuels such as petrol or diesel.

....

....

....

....

....

....

....

....

....

(6 marks)
(Total marks: 6)

Designing an investigation and making measurements

As well as demonstrating your investigative skills practically, you are likely to be asked to comment on investigations done by others. The example below offers guidance in this skill area. It also gives you the chance to practise using your skills to answer the types of question that may come up in exams.

Comparing the energy of different fuels

Skill – Understanding the experiment

> Josh tested the hypothesis that the bigger a fuel molecule, the more energy is released when it burns.
>
> He burnt 1 g of a fuel, methanol, and used it to heat 100 cm³ of water.
>
> Josh repeated the procedure with four other fuels. He used the apparatus below.
> Josh's results are in the table.

Fuel	Number of carbon atoms in one molecule	Temp before heating (°C)	Temp after heating (°C)	Temp change (°C)
Methanol	1	19	49	30
Ethanol	2	19	54	35
Propanol	3	20	61	41
Butanol	4	20	63	43
Pentanol	5	21	66	45

> 1 Identify the independent and dependent variables.

In an investigation:
- The independent variable is the one that is changed by the scientist or student.
- The dependent variable is the one that is measured for each change in the independent variable.

Try remembering it like this – the dependent variable *depends* on what you do to the independent one.

> 2 Identify three control variables.

In a fair test, only the independent variable should affect the dependent variable. The other variables must be kept the same. These are the control variables.

In Josh's investigation, one of the control variables is the volume of water. Can you identify the others?

Skill – Analysing the experiment

> 3 The smallest temperature change that Josh's thermometer can detect is 1 °C. Describe an advantage of using a thermometer that can detect a temperature change of 0.5 °C.

This question is asking about the resolution of the thermometer. In general, using a measuring instrument with a high resolution gives more accurate values (ones that are closer to the true value) than instruments with a lower resolution.

Skill – Evaluating the experiment

> 4 Do Josh's results support his hypothesis? Give a reason for your answer.

A hypothesis is an idea to explain observations. Josh uses the idea of increasing molecule size to explain why different fuels release different amounts of energy on burning. Do you think Josh's idea is correct? Why?

Skill – Using data to draw conclusions

> 5 Predict the temperature change of 100 cm³ of water if Josh burns 1 g of hexanol. Hexanol has six carbon atoms in one molecule.

A prediction is a statement about the way something will happen in the future. It is not the same as a hypothesis.

To answer this question properly, you would need to draw a graph or bar chart. But you can use the data in the table to make a rough prediction.

> 6 Rebecca does a similar investigation to Josh but she repeats the test for each fuel three times. The table below summarises her results.
> Suggest two advantages of repeating the test for each fuel.

Fuel	Temperature change (°C)			
	Run 1	Run 2	Run 3	Mean
Methanol	28	32	30	30
Ethanol	39	35	37	37
Propanol	43	61	43	43
Butanol	45	45	45	45
Pentanol	48	46	47	47

This question is asking about repeatability – how similar are the measurements when repeated under the same conditions by the same person?

If repeated values are close to each other, then the results are likely to be close to the true value.

You could also mention the anomalous result. Which result is anomalous? How does repeating tests help to identify anomalous data? Why might Rebecca have chosen to ignore the anomalous result?

AQA *Upgrade*

Answering an extended writing question

QUESTION

1 *In this question you will be assessed on using good English, organising information clearly, and using specialist terms where appropriate.*

Describe three ways by which copper metal can be obtained. Include descriptions of methods for extracting copper from its ores, and by recycling. Give an advantage or disadvantage of each method.

(6 marks)

---

you can heat copper ores this use emergie and you can use plants this is good and recycling is good two

G–E

Examiner: This answer is typical of a grade G candidate. It is worth just one mark.

The candidate has mentioned only two methods for obtaining copper metal, and has implied only one disadvantage of one method.

The candidate has not used specialist terms. There is no punctuation, and there are errors of grammar and spelling.

You can heat copper ores in a furnness. This is bad because it uses lots of energy. You can use bioleaching. This needs bacterea. It is good because it gets copper from ores with just a teeny bit of copper in them.

D–C

Examiner: This answer is worth three marks out of six. It is typical of a grade C or D candidate. The candidate has described two methods of extracting copper from ores, and has correctly given an advantage or disadvantage for each. No mention has been made of copper recycling.

The answer is well organised, with correct grammar and punctuation. There are a few spelling mistakes.

---

1 Copper can be obtained from copper-rich ores by heating them in a furnace. This is smelting. Then the copper is made pure by electrolysis. One disadvantage of this method is that it needs electricity. If the electricity is generated by burning fossil fuels, then the process puts carbon dioxide into the air, causing global warming.

Copper can be obtained from low-grade ores by phytomining. This uses plants to take in metal compounds. Then they burn the plants and get the copper compounds out of the ash. The advantage is that its energy costs are low, and it makes little solid waste.

You can also get copper by using waste copper to make solutions of salts. Then you add scrap steel, which is mainly iron. There is a displacement reaction, which gives copper.

$$\text{copper sulfate} + \text{iron} \rightarrow \text{iron sulfate} + \text{copper}$$

This is recycling. Its advantage is that it doesn't use up copper ores, and the process does not make greenhouse gases.

B–A*

Examiner: This is a high-quality answer, typical of an A* candidate. It is worth six marks out of six.

All parts of the question have been answered accurately and in detail.

The answer is well organised, and the spelling, punctuation, and grammar are faultless. The candidate has used several specialist terms.

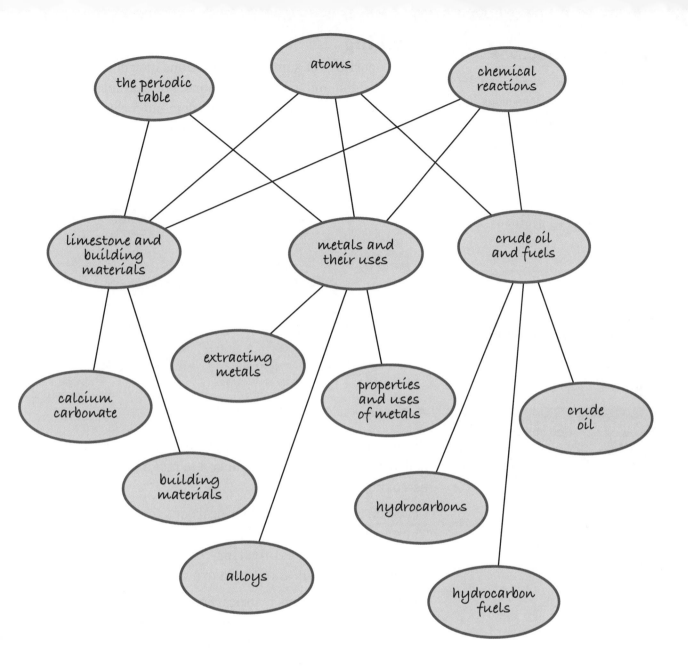

Revision checklist

- All substances are made up of atoms. Atoms are made up of protons, neutrons, and electrons.
- Elements are substances made up of one type of atom.
- There are about 100 elements listed in the periodic table.
- In the periodic table, elements in the same group have the same number of electrons in their highest energy level, and so have similar properties.
- In chemical reactions, atoms are rearranged to make new products. The mass of products equals the mass of the reactants.
- Limestone is a raw material for the manufacture of cement and concrete.
- Limestone is mainly calcium carbonate.
- Some metal carbonates decompose on heating to give carbon dioxide and a metal oxide.
- Carbonates react with acids to produce carbon dioxide, a salt, and water.

- Most metals are found combined with other elements in ores.
- Metals that are less reactive than carbon can be extracted from their ores by reduction with carbon.
- Metals that are more reactive than carbon are extracted by electrolysis of their molten compounds.
- An alloy is a mixture of a metal with small amounts of other elements. Many alloys are harder than the metals from which they are made.
- The uses of metals and alloys depend on their properties.
- Crude oil is a mixture of many hydrocarbons. Most of these are saturated hydrocarbons called alkanes.
- Fractional distillation separates the hydrocarbons in crude oil into fractions, each of which contains molecules with a similar number of carbon atoms.
- Hydrocarbon fuels burn to release energy. Some of their combustion products are pollutants.

Revision objectives

- ✓ identify an advantage of cracking crude oil fractions
- ✓ describe the process of cracking, and identify the conditions needed
- ✓ interpret the formulae of alkenes
- ✓ describe a test to identify alkenes

Student book references

1.18 Cracking crude oil

Specification key

✓ C1.5.1

Cracking

The proportions of some of the fractions in crude oil do not match the demand for these fractions. For example, the pie charts below show that the relative demand for petrol is much greater than its relative amount in crude oil.

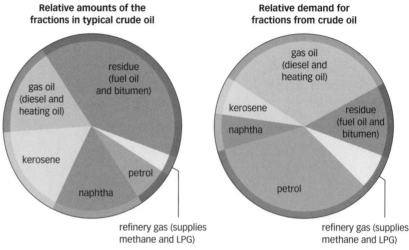

▲ The supply and demand for crude oil fractions.

Because of the mismatch in supply and demand, some hydrocarbon fractions are **cracked** to make smaller, more useful molecules.

Cracking involves heating hydrocarbons to vapourise them. The vapours can then be:

- passed over a hot **catalyst**, or
- mixed with steam and heated to a very high temperature.

Under these conditions hydrocarbon molecules in the vapour mixture break down to form smaller molecules in **thermal decomposition** reactions. The catalyst speeds up thermal decomposition reactions but is not itself used up.

Exam tip

When describing chemical tests, don't forget to describe the expected appearance of the test solution both before and after adding the substance being tested.

For example, on cracking the naphtha fraction, octane molecules may break down to make two smaller molecules – hexane and ethane.

$$octane \rightarrow hexane + ethene$$

H $\qquad C_8H_{18} \rightarrow C_6H_{14} + C_2H_4$

Hexane is a useful fuel. It is added to the petrol fraction. Other products of cracking are also useful fuels.

Alkenes

Ethene, one product of cracking, has a double bond between its two carbon atoms. This means it is an **unsaturated hydrocarbon**. It is a member of a group of hydrocarbons called the **alkenes**.

The table shows two alkenes.

Name	Molecular formula	Structural formula
Ethene	C_2H_4	
Propene	C_3H_6	

The alkenes have the general formula C_nH_{2n}. This shows that the number of hydrogen atoms in an alkene molecule is double the number of carbon atoms.

Detecting alkenes

You can detect unsaturated compounds by testing with **bromine water**. Orange bromine water becomes colourless when it reacts with alkenes.

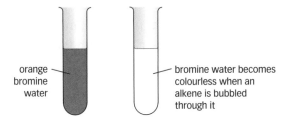

orange bromine water

bromine water becomes colourless when an alkene is bubbled through it

Key words

cracking, catalyst, thermal decomposition, unsaturated hydrocarbon, alkene, ethene, propene, bromine water

Exam tip

Make sure you can explain both *why* and *how* hydrocarbon fractions are cracked. Practise writing both types of formulae for ethene and propene.

Questions

1 Give one advantage of cracking the naphtha fraction of crude oil.

2 Give the conditions needed for cracking. Identify the two types of product of a typical cracking reaction.

3 Write the molecular formulae of ethene and propene.

4 Describe the test for identifying alkenes.

Revision objectives

- ✓ describe how polymers are made
- ✓ write equations to represent polymerisation reactions
- ✓ evaluate the impacts of the uses, disposal, and recycling of polymers

Making polymers

Alkenes are important raw materials. They are used to make **polymers**. Polymers are materials that have very big molecules. They are made by joining together many small molecules, called **monomers**. For example:

part of a poly(ethene) molecule

▲ Ethene molecules join together in long chains to make poly(ethene).

Using polymers

There are thousands of different polymers. Each has unique properties. The uses of a polymer depend on its properties. Scientists continue to develop new polymers with new applications.

Type of polymer	Properties	Uses
'Breathable' polymers	Allow water vapour to pass through their tiny pores, but not liquid water.	Making waterproof clothes.
Dental polymers	White, hard, and tough. Poor conductors of heat.	Fillings for teeth.
Hydrogels	Absorb huge volumes of liquid.	Making wound dressings and disposable nappies.
Smart materials, for example, shape-memory polymers	Change in response to their environment.	Making shrink-wrap packaging and glasses frames.

The properties of some polymers make them better for particular purposes than the material they replace. For example, metal fillings conduct heat well, making it uncomfortable to eat very hot or very cold food. New dental polymers are poor conductors of heat.

Polymer disposal

Many polymers are **non-biodegradable** – they cannot be broken down by microbes. This creates disposal problems because when put into **landfill sites** they remain unchanged for many years. When thrown away as litter, they persist in the environment for a long time, and may injure animals.

Chemists are developing ways of solving these problems, including:

- Recycling plastics – mixed plastic waste is collected and sorted into separate types of polymer, often by hand. This process is time consuming and expensive. The separate polymers are then melted and made into new products.

PET	high-density poly(ethene)	PVC	low-density poly(ethene)	poly(propene)	polystyrene

| PET | HDPE | PVC | LDPE | PP | PS |

▲ Recycling symbols found on plastic items.

- Adding starch to poly(ethene) – bacteria break down the starch once the polymer gets wet. This makes the plastic item break down into very small pieces. It has not rotted away, but it is no longer litter.
- Making bags from cornstarch – cornstarch is made from maize. It is biodegradable. There are advantages and disadvantages of making bags from cornstarch.

	Advantages	Disadvantages
Environmental	Biodegradable.	Fertiliser used to help maize grow may pollute rivers and lakes.
Economic	Growers and manufacturers have products to sell.	Cornstarch bags can be more expensive than poly(ethene) ones.
Social and ethical	Cornstarch is renewable, so using it to make bags does not take supplies from future generations.	Grown on land that could be used for food crops.

Key words

polymer, monomer, non-biodegradable, landfill site

Exam tip

Practise writing equations to represent polymerisation reactions.

Questions

1 Describe the properties of dental polymers, and explain why these properties make them suitable for their purpose.

2 Use ideas about properties to explain why hydrogels are used in disposable nappies.

3 Identify the environmental, economic, and social advantages of making bags from cornstarch instead of from poly(ethene).

Revision objectives

- describe two methods of manufacturing ethanol
- evaluate the advantages and disadvantages of the two methods

Student book references

1.23 Making ethanol

Specification key

✓ C1.5.3

Key words

hydration reaction, fermentation, enzyme

Exam tip

Remember – ethene is a non-renewable resource, but glucose is renewable.

Questions

1 Create a table to show the raw materials, conditions, and products of the two methods of manufacturing ethanol.

2 Evaluate the advantages and disadvantages of making ethanol from renewable and non-renewable resources.

Using ethanol

People use lots of ethanol – as a solvent, as a disinfectant, as a fuel, and in alcoholic drinks. So the demand for ethanol is high.

There are two ways of making ethanol. Each has its pros and cons.

Ethanol from ethene

In this method, ethene reacts with steam in the presence of a catalyst. Ethanol is the product of this **hydration reaction**.

$$\text{ethene} + \text{steam} \rightarrow \text{ethanol}$$

H $\qquad C_2H_4 \;+\; H_2O \;\rightarrow C_2H_5OH$

The reaction works well at about 300 °C

Ethanol from sugars

Ethanol can also be produced by **fermentation**. In fermentation, **enzymes** (natural catalysts) in yeast break down plant sugars into ethanol and carbon dioxide. For example:

$$\text{glucose} \rightarrow \text{ethanol} + \text{carbon dioxide}$$

H $\qquad C_6H_{12}O_6 \rightarrow 2C_2H_5OH + \qquad 2CO_2$

The reaction works best at about 37 °C.

Comparing the two ways of producing ethanol

	From ethene by hydration	From plant materials by fermentation
Raw materials	Ethene is produced from crude oil, which is non-renewable.	Plant materials are renewable.
Energy costs	Higher – both the cracking reaction to produce ethene, and the hydration reaction take place at high temperatures.	Lower – fermentation takes place at a lower temperature but the crops may require input such as fertiliser.

1 Tick the boxes to show which of the substances listed below are produced from crude oil:
poly(ethene) ☐ diesel ☐
petrol ☐ wool ☐
paper ☐

2 Draw lines to match each polymer use to the properties that make it suitable for this use.

Polymer use	Property
Dental fillings.	Allows water vapour to pass through its tiny pores, but not liquid water.
To make disposable nappies.	Hard and tough and a poor conductor of heat.
To make 'breathable' waterproof fabrics.	Changes shape in response to warming or pressure.
To make mattresses that mould to the body.	Absorbs large volumes of liquid.

3 Highlight the statements below that are true. Then write corrected versions of the statements that are false.
  a Monomers are small molecules that join together to form polymers.
  b A molecule of poly(ethene) is made by joining together thousands of ethane molecules.
  c Polymers have very big molecules.
  d The monomer propene makes poly(propane).

4 Choose words from the box below to fill in the gaps in the sentences that follow. The words in the box may be used once, more than once, or not at all.

> decomposition **bigger** **smaller** **catalyst** **condense** **steam** **vapourise** **vapours** **liquids** **combustion**

Hydrocarbons are cracked to make substances with _____ molecules. The process involves heating the hydrocarbons to _____ them. Next the _____ are passed over a hot _____, or mixed with _____ and heated to a very high temperature. Thermal _____ reactions then occur.

5 Draw lines to match each name below to a molecular formula and a structural formula.

Molecular formula	Name	Structural formula
C_2H_4	Ethene	
C_3H_6	Propene	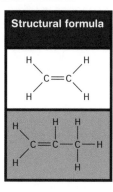

6 Alkenes have the general formula C_nH_{2n}. Tick the boxes to show which of the hydrocarbons below are alkenes.
C_4H_8 ☐ C_5H_{12} ☐
$C_{12}H_{26}$ ☐ C_7H_{14} ☐

7 A student bubbles four hydrocarbons through bromine water. Complete this table to show his expected observations.

Name of hydrocarbon	Results
Ethane	
Propene	
Ethene	
Butane	

8 Complete the equations below to show how poly(ethene) and poly(propene) are made from their monomers.

$$n \quad \begin{array}{c} H \\ \end{array} C = C \begin{array}{c} H \\ \end{array} \longrightarrow \left(\begin{array}{c} \\ C - C \\ \\ \end{array} \right)_n$$

ethene

$$n \quad \begin{array}{c} H \\ \end{array} C = C \begin{array}{c} H \\ CH_3 \end{array} \longrightarrow$$

9 Oil companies often crack the naphtha fraction in crude oil. The process produces hydrocarbons in the petrol fraction. Use data from these pie charts to suggest how cracking might benefit oil companies.

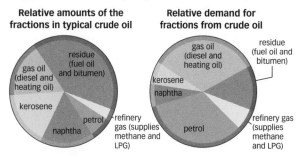

Relative amounts of the fractions in typical crude oil

Relative demand for fractions from crude oil

10 Complete this table to summarise the two ways of manufacturing ethanol.

Starting materials	Type of reaction	Conditions
Glucose		
		needs catalyst

11 Complete this table to show the environmental advantages and disadvantages of making ethanol from ethene and from glucose.

	Advantages	Disadvantages
From **ethene**		
From **glucose**		

12 A city council is deciding whether or not to provide a plastic recycling collection to all homes in the city. Describe the benefits and problems of recycling polymer waste.

1 This diagram shows the apparatus that can be used to crack a hydrocarbon in the laboratory.

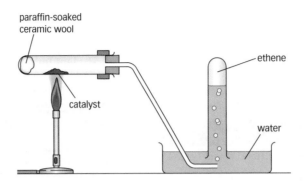

a What type of reaction occurs in the boiling tube?

Tick the **one** correct box.

Type of reaction	Tick
Reduction	
Oxidation	
Combustion	
Thermal decomposition	

(1 mark)

b The aluminium oxide shown in the apparatus above is not used up in the reaction. What is its purpose?

....
(1 mark)

c One product of the cracking reaction is ethene.

The displayed formula of ethene is

$$\begin{array}{c}H\\ \diagdown\\ C = C\\ \diagup \diagdown\\ H H\end{array}$$

i Give the molecular formula of ethene.

....
(1 mark)

ii Describe a chemical test to show that ethene has a double bond. Include the name of the chemical you need, brief instructions for doing the test, and the changes you would expect to see.

....

....

....
(3 marks)
(Total marks: 6)

2 In 2002, Bangladesh became one of the first countries to ban the use of all polythene bags in its capital city. The bags were found to have blocked the city's drainage system, which helped to cause severe flooding in 1988 and 1998.

a Give a scientific reason to explain why some polythene bags remained in drainage ditches for many years.

....

(1 mark)

b Today, many bags in Bangladesh are made from a plant material, jute. Suggest two advantages of making bags from jute compared to making them from polythene.

1 ....

2 ....

(2 marks)

(Total marks: 3)

3 Read the information in the box below, then answer the questions that follow.

> Poly(butene) is a polymer. It is flexible and withstands large forces without being damaged. It is slightly elastic (stretchy). It does not react with detergents, oils, fats, acids, bases, alcohols, or hot water. Over time, it may react with chlorine and its compounds.
>
> Poly(butene) has been used to make water pipes since the early 1970s. In Vienna, Austria, hot water from underground travels through poly(butene) pipes to heat homes. In the UK, many water pipes are made from poly(butene).

a Complete the equation below to show how poly(butene) is made from its monomer. Include the structural formula of poly(butene).

$$n \quad \overset{H}{\underset{H}{\diagdown}} C = C \overset{H}{\underset{C_2H_5}{\diagup}} \longrightarrow$$

(3 marks)

b **i** Explain why poly(butene) is a suitable material from which to make water pipes.

....

....

(2 marks)

ii In Canada, poly(butene) pipes are no longer used to transport drinking water to which chlorine and its compounds have been added. Suggest why.

....

....

(2 marks)

(Total marks: 7)

Revision objectives

- outline how vegetable oils are extracted from plant materials
- describe the effects of using vegetable oils in food and the impacts on diet and health

Student book references

1.24 Using plant oils
1.25 Oils from fruit and seeds

Specification key

✔ C1.6.1

Vegetable oils

The fruit, seeds, and nuts of some plants are rich in natural oils. These oils are called **vegetable oils**.

Extracting vegetable oils

Oils are extracted from different plants in different ways:

- Olive oil is extracted by crushing and **pressing** olives. The oil is then separated from water and other impurities.
- The flow diagram below shows how sunflower oil is extracted from sunflower seeds.

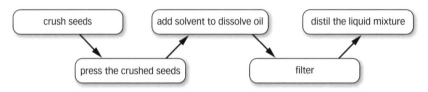

crush seeds → press the crushed seeds → add solvent to dissolve oil → filter → distil the liquid mixture

▲ Extracting sunflower oil from sunflower seeds.

- Lavender oil is extracted by **steam distillation**. The process produces a mixture of water and lavender oil. Water is removed from the mixture, and pure lavender oil remains.

Vegetable oils as food

Vegetable oils are important foods. They provide **nutrients**, such as vitamin E. They also release a lot of energy when digested, so eating large amounts of oil-rich foods can make people gain weight.

Vegetable oils boil at higher temperatures than water. So oils can cook foods at higher temperatures than boiling water. This means that:

- food cooks more quickly in oil
- a food cooked in oil tastes different to the same food cooked in water.

Vegetable oils as fuels

Since vegetable oils release large amounts of energy on burning, they are useful as vehicle fuels.

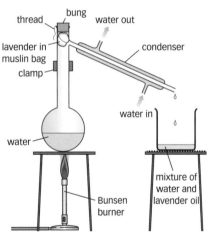

thread, bung, water out, lavender in muslin bag, condenser, clamp, water in, water, mixture of water and lavender oil, Bunsen burner

▲ Steam distillation is used to extract lavender oil. Note: in this diagram, the apparatus has been simplified.

Key words

vegetable oil, pressing, steam distillation, nutrient

Questions

1 Name three plant parts from which oils may be extracted.

2 Describe how olive oil is extracted from olives.

3 Explain why vegetable oils can be used as vehicle fuels.

Emulsions

You cannot mix vegetable oils with water. This is because their particles are too different. Vegetable oil molecules include long hydrocarbon chains. These chains cannot interact with small water molecules.

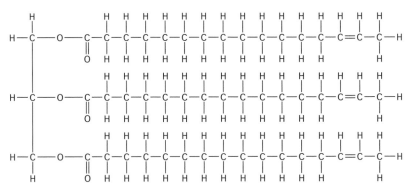

▲ A typical vegetable oil molecule.

But if you add an **emulsifier** to oil and water and shake well, the oil and water no longer separate out. The emulsifier stabilises the mixture. An **emulsion** forms.

Emulsions are thicker than oil or water. Their uses depend on their properties. Emulsions include:
- salad dressing – its thickness means it coats salad well
- icecream – its texture helps make it enjoyable to eat
- paints – their thickness means they coat walls well
- cosmetics such as hand creams, shaving cream, and sun screens

Revision objectives

- explain what emulsions are
- explain how emulsifiers stabilise emulsions
- describe how unsaturated vegetable oils can be hardened

Student book references

1.26 Emulsions

1.27 Making margarine

Specification key

✔ C1.6.2, ✔ C1.6.3

H Emulsifiers

Emulsifiers include egg – this prevents the oil and vinegar in mayonnaise separating. Emulsifier molecules have two ends:
- One end interacts with water molecules. This is the **hydrophilic end**.
- One end interacts well with oil molecules, and does not interact with water molecules. This is the **hydrophobic end**.

This diagram shows how emulsifier molecules stop an oil droplet separating from water.

Emulsifier safety

Egg is a natural emulsifier. Some foods contain artificial emulsifiers. These are identified by **E-numbers**. Additives with E-numbers have been safety tested and licensed by the European Union.

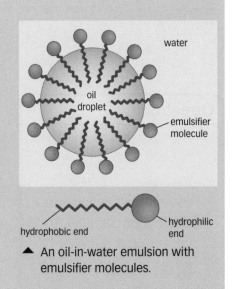

▲ An oil-in-water emulsion with emulsifier molecules.

Unsaturated oils

Sunflower oil is **unsaturated**. It contains double bonds between some of the carbon atoms in its molecules. You can represent a carbon–carbon double bond like this:

$$C = C$$

You can use bromine water to detect carbon–carbon double bonds.

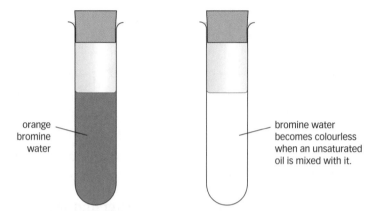

orange bromine water

bromine water becomes colourless when an unsaturated oil is mixed with it.

▲ Orange bromine water becomes colourless when shaken with an unsaturated vegetable oil.

Coconut oil is **saturated**. Its molecules do not have double bonds between their carbon atoms. Saturated fats raise blood cholesterol and increase the risk of heart disease. Unsaturated fats are better for health.

Exam tip   AQA

Remember the difference between saturated and unsaturated fats, and how to test for unsaturation.

Questions

1   Explain what an emulsion is.

2   Describe how to test a vegetable oil to find out whether or not it is unsaturated.

3   H Describe how unsaturated vegetable oils can be hardened.

H Hardening vegetable oils

Most vegetable oils are liquid at room temperature. Food companies harden unsaturated vegetable oils by adding hydrogen gas to them. The conditions for the reaction are:

- a temperature of 60 °C
- a nickel catalyst.

In these **hydrogenation** reactions, hydrogen atoms add to the carbon atoms on both sides of a double bond:

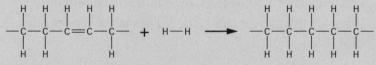

▲ This diagram shows just part of a plant oil molecule.

Hydrogenation reactions convert unsaturated oils into **saturated** ones. Saturated fats have higher melting points that unsaturated ones. So they are useful as spreads and for making cakes and pastries.

Working to Grade E

1 Tick the boxes to show the plant parts from which oils may be extracted.

seeds ☐   roots ☐

stem ☐   fruit ☐

nuts ☐

2 The statements below describe the steps in extracting sunflower oil from sunflower seeds. Write the letters of the steps in the best order.

a   Add a solvent.   c   Distil the mixture.

b   Crush the seeds.   d   Press the crushed seeds.

3 Highlight the statements below that are true. Then write corrected versions of the statements that are false.

a   Oils dissolve well in water.

b   A vegetable oil and water can be used to make an emulsion.

c   Emulsifiers make emulsions less stable.

d   A student shakes a mixture of oil, vinegar, and sugar. Afterwards, he sees two layers of liquid. This shows that sugar is an emulsifier.

4 This diagram shows the laboratory apparatus used to extract lavender oil from lavender plants.

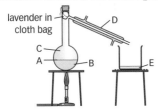

Answer each question below with the letter A, B, C, D or E. You may use each letter once, more than once, or not at all.

a   Where in the apparatus is a gas condensing?

b   Where is a liquid evaporating?

c   Which piece of apparatus contains water and lavender oil?

d   Which piece of apparatus contains tap water?

Working to Grade C

5 A student adds four plant oils to separate samples of bromine water. Her results are in the table below.

Name of oil	Observations on adding the oil to bromine water
Sunflower oil	Colour change from orange to colourless.
Coconut oil	Orange colour of bromine water is unchanged.
Palm oil	Orange colour of bromine water is unchanged.
Olive oil	Colour change from orange to colourless.

a   Name the oils that are saturated.

b   Name the oils that contain no double bonds between carbon atoms (C = C bonds).

6 Catherine and Sarah cook potatoes. Each uses a different cooking method. The table shows the energy released when they eat the potatoes.

Cook	Energy released per 100 g (kJ)
Catherine	1012
Sarah	288

a   Who cooked the potatoes in oil? Give a reason for your decision.

b   Give two advantages of cooking potatoes in oil, compared to boiling in water.

c   Give one disadvantage of cooking potatoes in oil, compared to boiling in water.

7 a   A doctor tells Alan to reduce the total amount of fat that he eats. Suggest why.

b   A doctor tells Jim to avoid saturated fats, and to eat unsaturated oils instead. Suggest why.

8 A student does an investigation to compare the energy released by three plant oils. He uses the apparatus below.

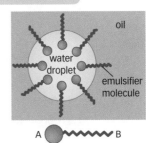

a   Identify the dependent and independent variables for the investigation.

b   Identify two control variables.

c   Explain how the student can use results from the investigation to work out which oil releases the most energy on burning.

Working to Grade A*

9 The diagram opposite shows a droplet of water surrounded by oil in an emulsion, and some emulsifier molecules. Which end of the emulsifier molecule is hydrophilic? Give a reason for your choice.

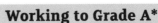

10 A company produces hardened oils from unsaturated vegetable oils.

a   Name the chemical that is added to unsaturated oils in order to harden them.

b   Give the conditions for the hardening process.

c   Describe one difference in the properties of an unsaturated oil and the hardened oil that is made from it.

d   Give two uses of hardened vegetable oils. Explain how their properties make them suitable for this use.

1 A student does an experiment to identify substances that act as emulsifiers.

- He places 3 cm³ of corn oil and 3 cm³ of water in a boiling tube.
- He adds a small amount of detergent.
- He puts a bung in the top of the boiling tube, and shakes.
- He repeats the experiment, using different substances instead of the detergent.

His results are given in the table below.

Name of substance	Observations
Detergent	Thick cream liquid. Can't see through it.
Salt	Two separate layers.
Mustard powder	Thick yellow liquid. Can't see through it.
Egg yolk	Thick yellow liquid. Can't see through it.
Flour	Two separate layers.

a Name the substances in the table that are emulsifiers.

....

(1 mark)

b i Salad dressing is an emulsion made from oil and vinegar. Explain why its properties make it suitable for use as a salad dressing.

....

(1 mark)

ii Name **two** other uses of emulsions.

....

(2 marks)

H

c This diagram shows a droplet of oil surrounded by water in an emulsion, and some emulsifier molecules.

Which end of an emulsifier molecule is hydrophobic?
Give a reason for your choice.

....

....

....

....

....

(2 marks)

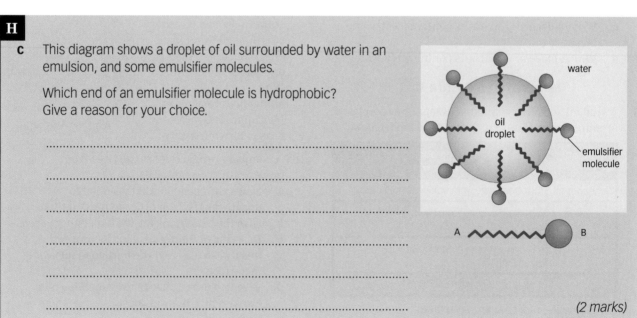

(Total marks: 6)

2 Describe **three** advantages and/or disadvantages of cooking potatoes in oil compared to cooking them in water.

....

....

....

(3 marks)

(Total marks: 3)

3 A teacher added bromine water to samples of four vegetable oils, then shook the mixtures. Her results are in the table below.

Vegetable oil sample	Observations
A	Orange bromine water became colourless.
B	Orange bromine water became colourless.
C	Orange colour did not change.
D	Orange colour did not change.

a i Give the letters of the **two** saturated vegetable oils in the table.

.... and ....

(1 mark)

ii Draw a ring around the one correct **bold** word or phrase in the sentence below.

Saturated vegetable oils have **zero/one/more than one** double bonds between carbon atoms.

(1 mark)

H

b i Describe and explain how unsaturated vegetable oils can be hardened to make hydrogenated vegetable oils.

....

....

....

(3 marks)

ii Give **two** uses of hydrogenated vegetable oils, and explain what makes them suitable for these uses.

1 ....

2 ....

(2 marks)

(Total marks: 7)

Revision objectives

- describe the structure of the Earth
- give evidence for Wegener's theory of crustal movement
- explain what tectonic plates are and how they move
- explain why it is not possible to predict exactly when earthquakes and volcanic eruptions will occur

Student book references

1.28 Inside the Earth

1.29 Moving continents

Specification key

- C1.7.1

Inside the Earth

The Earth consists of a **core**, **mantle**, and **crust**.

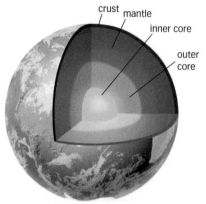

▲ The structure of the Earth.

Surrounding the Earth is a mixture of gases. This is the **atmosphere**. All the minerals and other resources that humans need come from the Earth's crust, the oceans, or the atmosphere.

Tectonic plates

The Earth's crust and upper part of the mantle are cracked into approximately 12 huge pieces, called **tectonic plates**.

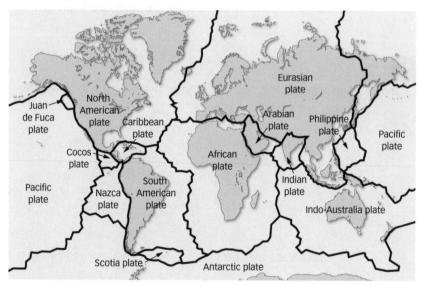

▲ Tectonic plates.

Tectonic plates are less dense than the part of the mantle that is below them, so they rest on top of it. The mantle is mostly solid but can move slowly.

Deep inside the Earth, natural radioactive processes heat the mantle. The heat drives convection currents within the mantle. These convection currents make tectonic plates move. They move slowly, at relative speeds of a few centimetres a year.

Earthquakes and volcanoes

Earthquakes and volcanic eruptions occur where tectonic plates meet.

- **Earthquakes** happen when tectonic plates move against each other suddenly. Scientists cannot predict when these movements will happen, so people living on plate boundaries can expect an earthquake at any time.
- Many **volcanoes** are at tectonic plate boundaries. Volcanoes often show signs of eruption in advance but the timing of an eruption depends on many factors. It is difficult to predict eruptions exactly.

Wegener's theory

For many years, some scientists thought that mountains and valleys were formed by the shrinking of the crust as the Earth cooled following its formation.

In 1912, Wegener put forward a new theory to explain the features of the Earth. He suggested that the continents were once joined together, and had gradually moved apart. Wegener supported his theory with evidence:

- The shapes of Africa and South America look as if they might once have fitted together.
- There are fossils of the same plants on both continents.
- There are rocks of the same type at the edges of the two continents where they might once have been joined.

At the time, most scientists did not accept Wegener's theory of **crustal movement**, or **continental drift**. This was because:
- Wegener was not a geologist
- they could not see *how* the continents might have moved.

Since the 1950s, scientists have discovered more evidence supporting Wegener's theory. Now his ideas are generally accepted.

Key words

core, mantle, crust, atmosphere, tectonic plate, earthquake, volcano, crustal movement, continental drift

Exam tip

Practise interpreting diagrams about Wegener's theory, and about the locations of earthquakes and volcanoes.

Questions

1 List the three sources of all the minerals and other resources that humans need.

2 At what speeds do tectonic plates move?

3 Explain what makes the Earth's mantle move.

4 Give three pieces of evidence to support Wegener's theory of crustal movement.

Revision objectives

- state the proportions of gases in the modern atmosphere
- describe one theory that explains how the early atmosphere was formed, and why the proportion of carbon dioxide decreased
- describe a scientific theory that explains how life began

Student book references

1.30 Gases in the air

1.31 Forming the atmosphere

Specification key

- C1.7.2 a–g

The modern atmosphere

For 200 million years, the proportions of gases in the atmosphere have been much the same as they are today.

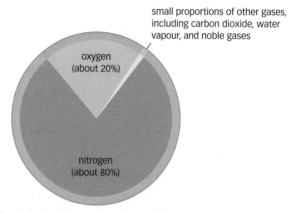

small proportions of other gases, including carbon dioxide, water vapour, and noble gases

oxygen (about 20%)

nitrogen (about 80%)

▲ This pie chart shows the proportions of the main gases in the air

Forming the atmosphere

During the first billion years of the Earth's existence, many volcanoes erupted. The gases from these eruptions formed the early atmosphere. Water vapour from the eruptions condensed to form the oceans.

There are several theories about how the atmosphere was formed. One theory suggests that the early atmosphere was mainly carbon dioxide gas. There were small amounts of water vapour, methane, and ammonia. There was little or no oxygen gas, like the modern atmospheres of Mars and Venus.

H Life begins

In this early atmosphere, life began. No one can be sure exactly how, or when, because no one was around to make observations at the time. There are several scientific theories to explain the origin of life.

One of these theories, the **primordial soup theory**, involves the interaction between:
- hydrocarbon compounds
- ammonia
- lightning.

According to the theory, gases in the early atmosphere reacted with each other in the presence of lightning, or sunlight, to make the complex molecules that are the basis of life.

In 1953, two scientists did an experiment to test the theory. The **Miller–Urey experiment** simulated a lightning spark in a mixture of the gases of the early atmosphere. A week later, more than 2% of the carbon in the system had formed compounds from which the proteins in living cells are made.

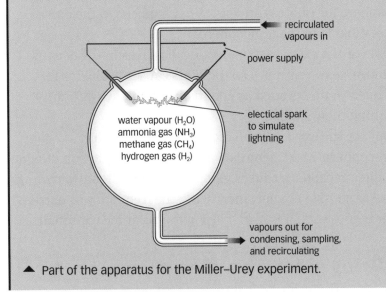

water vapour (H_2O)
ammonia gas (NH_3)
methane gas (CH_4)
hydrogen gas (H_2)

recirculated vapours in
power supply
electical spark to simulate lightning
vapours out for condensing, sampling, and recirculating

▲ Part of the apparatus for the Miller–Urey experiment.

Where did the oxygen come from?

Plants make their food by **photosynthesis**. They take carbon dioxide from the atmosphere and release oxygen gas.

carbon dioxide + water → glucose + oxygen

The photosynthesis of early plants and algae removed carbon dioxide from the atmosphere, and added oxygen gas to it.

What happened to the carbon dioxide?

Some of the carbon dioxide from the early atmosphere was removed by plants. But not all of it. What happened to the rest?

Locking up carbon in rocks

Much carbon dioxide dissolved in the oceans. Shellfish and other sea creatures used some of this dissolved carbon dioxide in making their shells and skeletons. The animals died and fell to the bottom of the sea. After many years, limestone, a **sedimentary rock**, formed from their shells and skeletons. The carbon atoms were locked away in the calcium carbonate of the limestone.

Locking up carbon in fossil fuels

Millions of years ago, dead animals and plants decayed under swamps. The dead organisms formed **fossil fuels**. The carbon atoms of the plants and animals were locked up in underground stores of coal, oil, and gas.

Exam tip

Remember that the proportion of oxygen in the atmosphere increased and the percentage of carbon dioxide decreased (until recently).

Questions

1 List the main gases in the modern atmosphere, and give their proportions.

2 Explain why the proportion of carbon dioxide in the atmosphere is less now than it was millions of years ago.

3 Explain why no one knows how life began.

Revision objectives

- describe some impacts of the recent increase in the proportion of carbon dioxide in the atmosphere
- name some products that we obtain from the air, and state how they are extracted

Student book references

1.32 The carbon cycle

Specification key

- C1.7.2 h–j

Exam tip

AQA

Remember the problems caused by adding extra carbon dioxide to the oceans and to the atmosphere.

Carbon dioxide – on the up

Today, humans burn fossil fuels. This puts extra carbon dioxide into the atmosphere. Humans also destroy forests, which mean that less carbon dioxide is removed from the atmosphere for photosynthesis. What happens to the extra carbon dioxide in the atmosphere?

- Some of the extra carbon dioxide dissolves in the oceans, making seawater more acidic. This causes problems for living organisms such as shellfish, which have difficulty making their shells.
- Some of the extra carbon dioxide remains in the atmosphere. Most scientists agree that extra carbon dioxide in the atmosphere causes global warming. Global warming causes climate change, increasing the frequency of extreme weather events. It is also making the polar ice caps melt.

H Using gases from the air

The gases in the air have different boiling points. This means that the gases can be separated by fractional distillation. The fractional distillation of air provides:

- nitrogen gas, which is used as a raw material to make fertiliser, and to freeze food
- oxygen gas, for medical treatments
- noble gases, including argon and neon. Argon is used in double glazing. Neon is used in display lighting.

Questions

1   Explain why the oceans are becoming more acidic. Describe one problem this increasing acidity may cause.

2   H List three gases that are obtained by the fractional distillation of air, and give a use of each gas.

1 Label the different parts of the Earth's structure. Use these words: **mantle, crust, core**.

2 In each sentence below, highlight the correct **bold** word or phrase.

  a Tectonic plates consist of huge slabs of the Earth's **crust/crust and upper part of the mantle**.

  b The mantle is mainly **solid/liquid**.

  c The movement of tectonic plates is caused by **convection currents/ocean currents**.

  d The currents that make tectonic plates move are driven by natural **combustion reactions/ radioactive processes**.

3 Complete the table below to show the percentage of each gas in the Earth's atmosphere today.

Gas	Percentage
Nitrogen	
	20
Carbon dioxide, water vapour, noble gases	

4 Complete the sentences below using the words in the box.

> **oxygen   algae   photosynthesis carbon dioxide**

Plants and _____ take in _____ gas from the atmosphere. They use this gas in a process called _____. As a result of this process, they produce _____ gas.

5 Tick the boxes next to sources from which humans obtain minerals and other resources.

  a Earth's crust ☐
  b Earth's core ☐
  c Earth's mantle ☐
  d oceans ☐
  e Earth's atmosphere ☐
  f atmosphere of Mars ☐

6 Highlight the sentences below that are true. Then write corrected versions of the statements that are false.

  a Burning fossil fuels increases the amount of carbon dioxide in the atmosphere.

  b Over the past few years, the amount of carbon dioxide absorbed by the oceans has decreased.

  c Most scientists agree that the decreasing amounts of carbon dioxide in the atmosphere are causing global warming.

  d The oceans act as a reservoir, or store, for carbon dioxide.

7 In 1912, Wegener proposed his theory of crustal movement, or continental drift.

  a List three pieces of evidence that Wegener used to support his theory.

  b Give two reasons to explain why other scientists did not at first support the theory.

8 During the first billion years of the Earth's existence, volcanoes released huge amounts of carbon dioxide into the atmosphere. Then, the percentage of carbon dioxide in the atmosphere gradually decreased. Complete the flow diagram below to summarise one theory that explains this decrease. Use the words and phrases in the box.

> **dissolving   photosynthesis decay in absence of oxygen sedimentary rock**

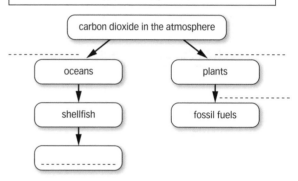

9 Describe the 'primordial soup' theory of the origin of life.

10 This diagram shows part of the apparatus for the Miller–Urey experiment, which was designed to test the 'primordial soup' theory.

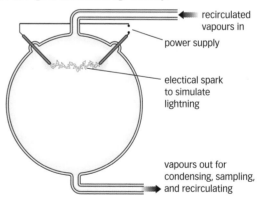

  a What does the electrical spark simulate?

  b Name two types of molecule that reacted together in the Miller–Urey experiment.

  c Explain why no-one knows whether or not the 'primordial soup' theory is correct.

11 a Name the process by which the gases of the air can be separated.

  b Give the names of two noble gases that are obtained from the air by this process, and give one use for each gas.

Examination questions
Changes in the Earth and its atmosphere

1 In February 2010, there was an earthquake off the coast of Chile. About 500 people died. This map shows the epicentre of the earthquake and the tectonic plates in the region.

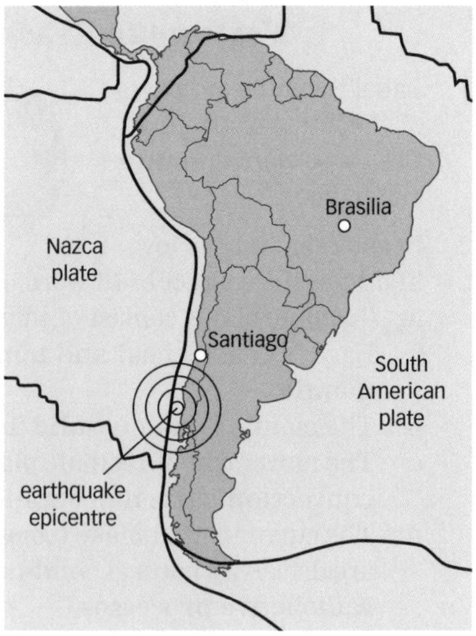

a What are tectonic plates made up of? Tick **one** box.

Tectonic plates are made up of . . .	Tick
the Earth's crust.	
the Earth's crust and the whole thickness of the mantle.	
the Earth's crust and the upper part of the mantle.	
the Earth's crust and the upper core.	

(1 mark)

b How fast do tectonic plates move, relative to each other? Tick **one** box.

Tectonic plates move at a speed of . . .	Tick
a few millimetres a year.	
a few centimetres a year.	
a few metres a year.	
a few kilometres a year.	

(1 mark)

c Explain what makes tectonic plates move.

....

....

(2 marks)

d Look at the map above.

Explain why there is a greater risk of earthquakes in Santiago than there is in Brasilia.

....

(1 mark)

e Explain why scientists cannot accurately predict when earthquakes will happen.

....

....

(1 mark)

(Total marks: 6)

2 This table shows the composition of the atmosphere of the planet Mars.

Gas	Percentage in atmosphere of Mars
Carbon dioxide	95
Nitrogen	3
Argon	1.6
Oxygen, methane, water vapour	Very small amounts

a i Sketch a line graph, bar chart, or pie chart to represent the information in the table. Make sure you sketch the most appropriate type of graph or chart.

(2 marks)

ii Use the information in the table, and your own knowledge, to describe one similarity and one difference between the modern atmospheres of Mars and the Earth.

Similarity: ....

Difference: ....

(2 marks)

b There are several theories about how the modern atmosphere of the Earth was formed. One theory suggests that the early atmosphere of the Earth was similar to that of Mars today.

i Where did the carbon dioxide of the Earth's early atmosphere come from?

....

(1 mark)

ii Describe **two** ways by which carbon dioxide was removed from the early atmosphere. Include scientific words in your answer.

....

....

(2 marks)

c The concentration of carbon dioxide in the atmosphere has been increasing for the past 150 years. Give **two** reasons for this increase.

....

....

(2 marks)

(Total marks: 9)

Societal aspects of scientific evidence

This module includes examples of scientific issues that are influenced by the views of a wide range of people and organisations. It also includes examples of the use of scientific knowledge to make technological advances.

You may be asked to link developments in science to ethical, social, economic, or environmental issues. The example below supports you in identifying, explaining, and evaluating these links

Using vegetable oils as vehicle fuels

Skill – Analysing the facts and making deductions

A government asks three scientists to investigate using a fuel made from oilseed rape to replace petrol and diesel in cars. The scientists compare the factors below:

- the energy released on burning the fuels
- the land areas needed to grow and process the oilseed rape, and to obtain petrol and diesel from crude oil
- the waste products made when the fuels burn in cars.

Scientists from several organisations do the research, including:

- a scientist working at an oil company
- a scientist employed by a company that processes oilseed rape plants to make fuel
- a university research scientist funded by an independent charity.

1 Which of the scientists listed may be biased? Give a reason for your choice(s).

To answer this question, you will need to think about who is paying for the research. Would the funder prefer one outcome to another? You also need to consider the organisations the scientists work for – might they profit if the results of the research show one thing rather than another?

There is no one right answer to this question; it is your reasoning that is important here.

2 Suggest one other factor that might influence which scientist's work the government takes most seriously.

Again, there is no one right answer. You could mention things like the scientists' qualifications, experience, or status.

Skill – Understanding the impact of a decision

3 Below are some people's opinions about the two types of fuel. Sort the opinions into four groups depending on whether they are using ethical arguments, environmental arguments, economic arguments, or social arguments.

Georgia – I think that oilseed rape fuel produces more particulate pollution on burning than petrol does.

Hari – It is not right to grow fuel crops on land that could be used to grow food for starving people.

Imogen – Growing and processing oilseed rape in order to make fuel could provide many new jobs.

Julia – Obtaining and processing fossil fuels to get petrol and diesel is very expensive, and is likely to get even more costly in future.

Krishnan – Oilseed rape crops remove carbon dioxide from the atmosphere as they grow. So this fuel contributes less to global warming.

Lydia – I work for an oil company. Our profits will decrease if oilseed rape fuel replaces petrol and diesel.

This part of the question is asking you to identify four types of impact of producing and using the fuels.

- **Environmental impacts** are effects on the surroundings in which plants and animals live.
- **Social impacts** are impacts on people, or groups of people.
- **Economic impacts** are to do with money.
- **Ethical impacts** are to do with morals – is an action right or wrong?

Skill – Applying scientific principles to problems

4 Look again at the opinions of Georgia, Hari, Imogen, Julia, Krishnan, and Lydia. Which of the opinions could be investigated scientifically? Give a reason for your answer.

Here, you need to distinguish between questions that can be answered by doing investigations or studying data, and those that science cannot answer directly.

AQA Upgrade

Answering an extended writing question

In this question you will be assessed on using good English, organising information clearly, and using specialist terms where appropriate.

Identify the benefits and drawbacks of using plant oils to produce vehicle fuels, compared to producing fuels from crude oil. In your answer, you should state whether each benefit and drawback relates to social, economic, or environmental issues.

(6 marks)

QUESTION

G–E

Plant oils are cheeper than crude oil – economic advantige! I think its becoz its easier to grow plants than get oil from under the north see!!!

Examiner: This answer is worth two marks out of six. It is typical of a grade F or G candidate.

The answer includes one economic advantage of plant oils compared to crude oil. There are several spelling and punctuation mistakes.

D–C

Plant oils put carbon dioxide into the atmosfere when you burn them, but the plants take it out wen they grow. This is enviromental.

You grow the plants on land but people are starveing so it wood be better to use the land for food. This is soshal drawback.

Plants oils are reknewable and you can grow them again. This is soshal benifit becoz our grandchilds want fuels for cars two.

Examiner: This answer is worth three marks out of six, and is typical of a grade C or D candidate.

Three impacts of using plant oils have been described, and correctly identified as social, economic, or environmental advantages or disadvantages. The impacts of using plant oils have not been compared to the impacts of using crude oil to produce fuels.

The answer is well organised. There are several spelling mistakes, including words that are given in the question.

B–A*

Some farmers grow plants to make oils for fuel. A social and ethical drawback of this is that the plants grow on land that could be used to grow food. Less land is needed to produce fuels from crude oil.

An environmental benefit is that when the plants grow, they take carbon dioxide from the atmosphere for photosynthesis. But at the same time the tractors for spreading fertilisers put carbon dioxide into the air because they burn diesel. This is an environmental drawback. Both sorts of fuel put carbon dioxide into the air when they burn.

Plants are renewable, but crude oil is not. This is a benefit of making fuels from plant oils.

You can transport both types of fuel easily because they are both liquids. This is a benefit of both.

Examiner: This answer gains six marks out of six, and is typical of an A or A* candidate. All parts of the question are answered, and each benefit or disadvantage is labelled as being a social, economic, or environmental issue.

The answer is logically organised, and includes scientific words. The spelling, punctuation, and grammar are accurate.

Polymers, plant oils, the Earth, and its atmosphere

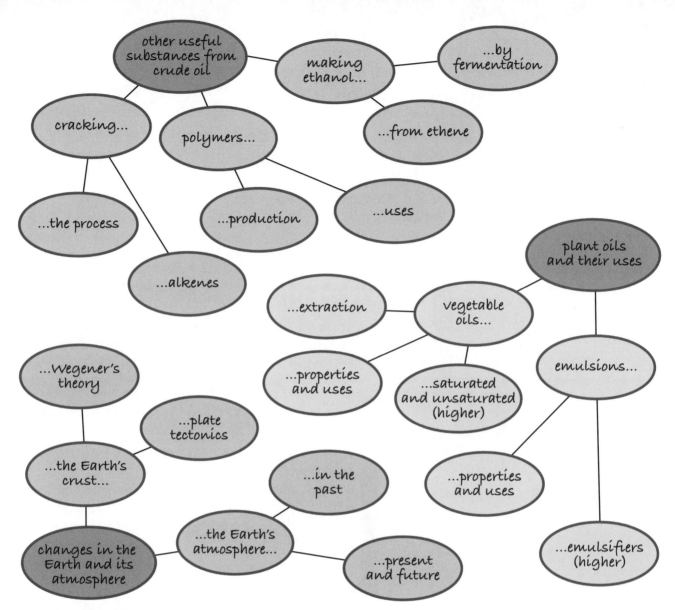

Revision checklist

- Hydrocarbons can be cracked to produce smaller and more useful molecules.
- Unsaturated hydrocarbons called alkenes are produced in cracking reactions.
- Small molecules (monomers) join together in polymerisation reactions to make big molecules (polymers).
- Scientists develop new polymers with properties for particular purposes.
- Ethanol is produced by the hydration of ethene or the fermentation of sugar.
- Vegetable oils are extracted by pressing or distillation from the seeds, fruits, and nuts of some plants.
- Vegetable oils boil at high temperatures and cook food quickly.
- Emulsions are made from vegetable oils and water. They are used in salad dressings, paint, and cosmetics.
- Emulsifiers make emulsions more stable.
- Their molecules have hydrophilic and hydrophobic ends.
- Unsaturated oils have double carbon–carbon bonds. They are better for health than saturated oils.

- Unsaturated oils are hardened by adding hydrogen in the presence of a catalyst at 60 °C.
- The Earth's crust and upper mantle are cracked into tectonic plates.
- Tectonic plates move relative to each other. Sudden movements cause earthquakes at plate boundaries.
- The Earth's atmosphere is about 80% nitrogen and 20% oxygen, with small proportions of carbon dioxide, water vapour, and noble gases.
- One theory suggests the early atmosphere was mainly carbon dioxide from volcanic eruptions.
- Carbon dioxide was removed from the early atmosphere by photosynthesising plants, and by dissolving in the oceans. Carbon from the carbon dioxide became locked up in sedimentary rocks and fossil fuels.
- Burning fossil fuels increases the amount of carbon dioxide in the atmosphere. This causes global warming.

Compounds

A **compound** is a substance that is made up of two or more elements. The atoms in a compound are held together by chemical bonds.

Chemical bonding involves the electrons in the highest occupied energy levels, or shells, of atoms. These electrons can be transferred from one atom to another or shared between two atoms.

By transferring or sharing electrons, atoms achieve the stable electronic structure of a noble gas (group 0 elements).

Making ions

When atoms transfer electrons, **ions** are made. An ion is an electrically charged atom, or group of atoms. Electrons are negatively charged, so:

- if an atom loses one or more electrons, it becomes a positive ion
- if an atom gains one or more electrons, it becomes a negative ion.

Most ions have 8 electrons in their highest energy level, like a noble gas atom.

A magnesium atom has 12 protons and 12 electrons. It loses two electrons to become a magnesium ion. The magnesium ion has 12 protons and 10 electrons. This gives it an overall charge of +2. Its formula is Mg^{2+}. You can represent its electronic structure as $[2,8]^{2+}$.

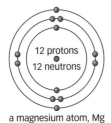
a magnesium atom, Mg

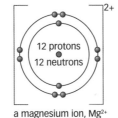
a magnesium ion, Mg^{2+}

An oxygen atom has 8 protons and 8 electrons. It gains two electrons to become an **oxide ion**. The oxide ion has 8 protons and 10 electrons. This gives it an overall charge of –2. Its formula is O^{2-}. Its electronic structure is $[2,8]^{2-}$.

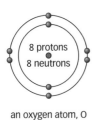
an oxygen atom, O

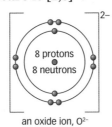
an oxide ion, O^{2-}

Magnesium oxide is made up of Mg^{2+} ions and O^{2-} ions. Its formula is MgO.

Revision objectives

- ✓ explain how ions form
- ✓ use dot-and-cross diagrams to represent ions
- ✓ write formulae for ionic compounds
- ✓ describe ionic bonding

Student book references

2.3 Ionic bonding
2.4 Making ionic compounds

Specification key

✓ C2.1.1 a – f

Key words

compound, ion, oxide ion, alkali metal, halogen, metal halide, ionic compound, giant ionic lattice, ionic bonding

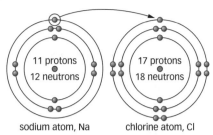

When sodium and chlorine react together, each sodium atom transfers one electron to a chlorine atom.

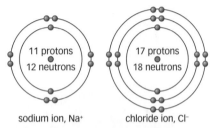

sodium ion, Na⁺   chloride ion, Cl⁻

▲ The diagrams show the electron arrangements in sodium and chloride ions.

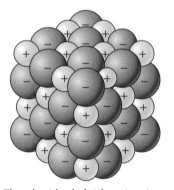

▲ The giant ionic lattice structure of sodium chloride.

Reactions of group 1 elements

The elements in group 1 of the periodic table are the **alkali metals**. They react vigorously with group 7 elements, the **halogens**. **Metal halides** are made. For example:

$$\text{sodium + chlorine} \rightarrow \text{sodium chloride}$$
$$2\text{Na} + \text{Cl}_2 \rightarrow 2\text{NaCl}$$
$$\text{lithium + bromine} \rightarrow \text{lithium bromide}$$
$$2\text{Li} + \text{Br}_2 \rightarrow 2\text{LiBr}$$
$$\text{potassium + iodine} \rightarrow \text{potassium iodide}$$
$$2\text{K} + \text{I}_2 \rightarrow 2\text{KI}$$

In these reactions, each metal atom transfers one electron to a halogen atom. This forms:

• metal ions with a single positive charge, such as K^+
• halide ions with a single negative charge, such as I^-.

Potassium iodide is made up of potassium ions, K^+ and iodide ions, I^-. Its formula is KI.

Group 1 metals also react vigorously with oxygen. For example:

$$\text{lithium + oxygen} \rightarrow \text{lithium oxide}$$
$$4\text{Li} + \text{O}_2 \rightarrow 2\text{Li}_2\text{O}$$

The products of the reactions of alkali metals with non-metal elements are all ionic compounds. In each compound the metal ion has a single positive charge.

Inside ionic compounds

Compounds made up of ions are **ionic compounds**. An ionic compound is a giant structure of ions. The ions are held together by strong electrostatic forces of attraction between the oppositely charged ions. These forces act in all directions. They hold the ions in a regular pattern called a **giant ionic lattice**. This is **ionic bonding**.

Questions

1   Draw dot-and-cross diagrams to represent the electronic structures of the ions in sodium chloride and calcium chloride.

2   Write a word equation for the reaction of sodium with bromine.

3   Describe the bonding in a crystal of sodium chloride.

Covalent bonds

In some elements and compounds, atoms share pairs of electrons. They do this to achieve the stable electronic structure of a noble gas. A shared pair of electrons is called a **covalent bond**. Covalent bonds are very strong.

Simple molecules

Some covalently bonded substances consist of simple molecules. A simple molecule is made up of a small number of atoms, with covalent bonds between the atoms. For example:

- Chlorine exists as chlorine molecules, Cl_2. A chlorine molecule consists of two chlorine atoms. Each chlorine atom shares one of its electrons with the other chlorine atom to make a covalent bond.

In this way, each chlorine atom has a share of eight electrons in its highest occupied energy level. The chlorine atoms now have the stable electronic structure of argon, a noble gas.

- Hydrogen exists as hydrogen molecules, H_2. Each hydrogen atom has the stable electronic structure of the noble gas helium.
- An oxygen molecule, O_2, consists of two oxygen atoms. Two pairs of electrons are shared between the atoms. The two shared pairs of electrons form a strong **double covalent bond**.

The atoms in the compounds in the table are joined together by covalent bonds to make simple molecules. Each line in a displayed formula represents a covalent bond.

Name of compound	Molecular formula	Dot-and-cross diagrams	Displayed formula
hydrogen chloride	HCl	H — Cl
water	H_2O	O—H / H
ammonia	NH_3	H—N—H / H
methane	CH_4	H—C—H

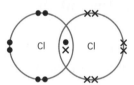

▲ This dot-and-cross diagram shows the electrons in the highest occupied energy levels of a chlorine molecule, Cl_2.

▲ This dot-and-cross diagram shows the electrons in a hydrogen molecule, H_2.

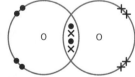

▲ Two pairs of electrons are shared between the atoms in an oxygen molecule.

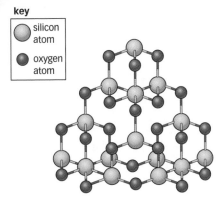

▲ The atoms in silicon dioxide are also joined together in a giant covalent structure.

Giant structures

The atoms in some covalently bonded substances are joined together in huge networks called **giant covalent structures**, or **macromolecules**.

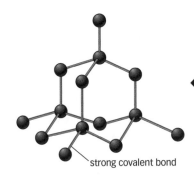

strong covalent bond

◀ Diamond is a form of the element carbon. In diamond, strong covalent bonds join each carbon atom to four other carbon atoms. The diagram shows just a tiny part of the structure of diamond.

Metallic bonding

Metals consist of giant structures of atoms arranged in a regular pattern.

Key words

covalent bond, simple molecule, double covalent bond, giant covalent structure, macromolecule, delocalised, giant metallic structure, metallic bonding

H In metals, the electrons in the highest energy levels of the atoms become **delocalised**. These electrons are no longer part of single atoms, but are free to move through the whole structure of the metal. The atoms that have lost electrons are positive ions. These are arranged in a regular pattern as a **giant metallic structure**. There are strong electrostatic forces of attraction between the positive ions and the moving delocalised electrons. This is **metallic bonding**.

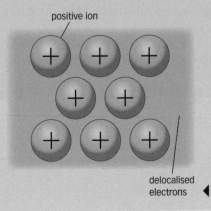

positive ion

delocalised electrons ◀ Metallic bonding.

Questions

1 Describe the differences between simple molecules and macromolecules.

2 Draw displayed formulae for molecules of hydrogen, oxygen, hydrogen chloride, and water.

3 **H** Describe the differences between the bonding in giant covalent and giant metallic structures.

Working to Grade E

1 Tick the boxes to show which substances listed below are compounds.
- **a** Oxygen
- **b** Carbon dioxide
- **c** Magnesium chloride
- **d** Chlorine
- **e** Germanium
- **f** Potassium bromide
- **g** Water

2 Highlight the statements below that are true. Then write corrected versions of the statements that are false.
- **a** A compound is made up of atoms of two or more elements.
- **b** When atoms share electrons, they form ionic bonds.
- **c** Diamond has a giant covalent structure.
- **d** Elements in group 1 form ions with a charge of +1.
- **e** Elements in group 7 form ions with a charge of +7.

3 Choose words from the box below to fill in the gaps in the sentences that follow. The words in the box may be used once, more than once, or not at all.

positively	noble	gas
molecules	metals	ions
negatively	alkali	seven
zero	one	eight

When atoms form chemical bonds by transferring electrons, they form _____. Atoms that lose electrons become _____ charged _____. Atoms that gain electrons become _____ charged ions. Ions have the electronic structure of a _____ _____, or group _____ atom.

4 Name the compounds formed when the following pairs of elements react together.
- **a** Sodium and chlorine.
- **b** Lithium and oxygen.
- **c** Potassium and iodine.
- **d** Sodium and bromine.

5 Draw lines to link each diagram to the type of structure that it represents.

Type of structure	Diagram
ionic	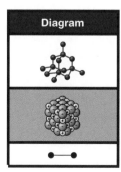
simple molecular	
giant covalent	

Working to Grade C

6 Tick the boxes to show which formulae below represent compounds.
- **a** Cl_2
- **b** Na
- **c** HCl
- **d** H_2
- **e** CH_4
- **f** O_2

7 Complete the table to show the numbers of protons and electrons in the ions listed. Use data from the periodic table to fill in the first empty column.

Ion	Number of protons	Number of electrons
Li^+		
F^-		
Na^+		
Cl^-		
Mg^{2+}		
Br^-		
Ca^{2+}		

8 Write the electronic structures of the ions below.
- **a** Na^+
- **b** Cl^-
- **c** Mg^{2+}
- **d** O^{2-}
- **e** Ca^{2+}
- **f** F^-

9 Give the formulae of the compounds that are made up of the ions listed in the table.

Formula of positive ion	Formula of negative ion	Formula of compound
Na^+	Cl^-	
Mg^{2+}	O^{2-}	
Ca^{2+}	Cl^-	
Rb^+	O^{2-}	

10 Draw dot-and-cross diagrams and displayed formulae to show the atoms and covalent bonds in the substances listed below.
- **a** H_2
- **b** Cl_2
- **c** O_2
- **d** HCl
- **e** H_2O
- **f** NH_3
- **g** CH_4

Working to Grade A*

11 The diagram below represents the bonding in a metal. Use the diagram to help you explain the structure of the metal, and how it is held together.

67

Examination questions
Structure and bonding

1 a Potassium chloride is an ionic compound.

i Complete the sentences below.

Potassium chloride is made up of two types of particle. These particles are positively charged potassium ions and .... charged .... ions.

(2 marks)

ii Use the periodic table to help you work out the number of protons in a potassium ion.

....

(1 mark)

iii Complete the diagram below to show all the electrons in a potassium ion.

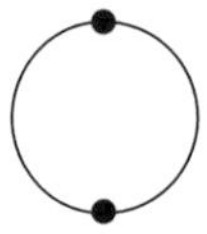

....

(2 marks)

iv Name and describe the forces that hold solid potassium chloride together.

....

....

....

(3 marks)

1 b A chemist makes potassium chloride in the laboratory from two starting materials: potassium and chlorine.

i Name the type of bonding in the element potassium.

....

(1 mark)

ii Name the type of bonding in a chlorine molecule.

....

(1 mark)

iii The diagram below shows the electrons in the highest occupied energy level (outer shell) of a chlorine atom.

Add to the diagram to show how the electrons are arranged in a chlorine molecule. Show the electrons in the highest occupied energy levels (outer shells) **only**.

(2 marks)
(Total marks: 12)

Student book references

2.6 Molecules and properties

2.7 Properties of ionic compounds

2.10 Metals

Specification key

✓ C2.2.1 ✓ C2.2.2 ✓ C2.2.4

Molecules

Many non-metal elements, and compounds made up of non-metals, consist of simple molecules. These substances:

- do not conduct electricity, because the molecules do not have an overall electric charge
- have relatively low melting points and boiling points.

H In substances that consist of simple molecules:

- the covalent bonds that join the atoms together in a molecule are strong
- the forces of attraction between each molecule and its neighbours – the **intermolecular forces** – are weak.

It is the weak intermolecular forces that must be overcome when a substance melts or boils, not the covalent bonds. This is why substances that consist of molecules have low melting and boiling points.

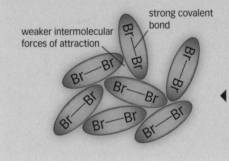

◀ This diagram shows the covalent bonds and intermolecular forces in liquid bromine.

Ionic compounds

Sodium chloride is an ionic compound. It is made up of positive sodium ions and negative chloride ions. These ions are arranged in a regular pattern called a **giant ionic lattice**.

There are very strong electrostatic forces of attraction between the sodium and chloride ions. The forces act in all directions.

Much energy is needed to break the strong bonds between the ions in sodium chloride, so sodium chloride has a high melting point and a high boiling point.

All ionic compounds have high melting points and boiling points because of the large amounts of energy needed to break their many strong bonds.

Solid ionic compounds do not conduct electricity. This is because their ions are not free to move from place to place to carry the current.

If you melt an ionic compound, or dissolve it in water, its ions become free to move. The free ions carry the current.

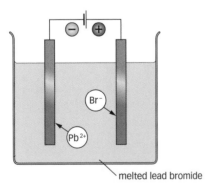

▲ Lead bromide is an ionic compound. Liquid lead bromide conducts electricity because its ions are free to move towards the electrodes.

Metals

H Metals conduct heat and electricity. This is because their delocalised electrons are free to move throughout the metal.

In metals, the layers of atoms can slide over each other easily. This is why it is easy to bend metals, and make them into different shapes.

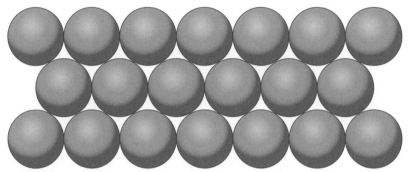

▲ The atoms in a metal.

An **alloy** is a mixture of a metal with one or more other elements. The other elements are usually metals.

The different elements in an alloy have atoms of different sizes. The different sized atoms distort the layers in the metal structure. This makes it more difficult for the layers to slide over each other. So alloys are harder than pure metals.

Shape-memory alloys such as Nitinol are used to make dental braces. They can go back to their original shape after being bent or twisted.

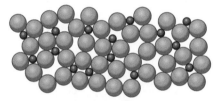

▲ An alloy is a mixture of a metal with small amounts of one or more other elements.

Questions

1   Explain why substances that consist of simple molecules do not conduct electricity.

2   Draw a table to compare the properties of ionic compounds, metals, and substances that consist of simple molecules.

3   **H** Explain why substances that consist of simple molecules have relatively low melting and boiling points.

Exam tip   AQA

You may be given the properties of a substance, and asked to suggest what type of structure it has. Practise doing this before the exam.

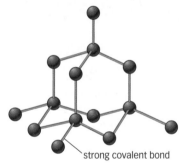

strong covalent bond

▲ Part of the structure of diamond.

Giant covalent structures

Some covalently bonded substances form giant structures, or macromolecules. For example:

- **Diamond** – a form of carbon.
- **Graphite** – another form of carbon.
- Silicon dioxide.

The atoms in these substances are arranged in huge repeating patterns, or **lattices**.

There are very strong covalent bonds between the atoms. Large amounts of energy are needed to break these bonds, so substances with giant covalent structures have very high melting points.

Diamond

In diamond, each carbon atom is joined to four other carbon atoms by strong covalent bonds. A giant covalent structure results. This explains why diamond is very hard.

Graphite

The carbon atoms in graphite have a different arrangement to those in diamond. This gives graphite and diamond different properties.

In graphite, each atom forms covalent bonds with three other atoms, making a layer. A lump of graphite consists of many of these layers. There are no covalent bonds between the layers, just weak intermolecular forces. This means the layers can slide over each other, making graphite soft and slippery.

H For every carbon atom of graphite, three of its four electrons in the highest energy level are involved in covalent bonding. The other electron in each carbon atom is delocalised, or free to move. The delocalised electrons in graphite can carry an electric current and help conduct heat.

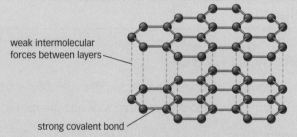

weak intermolecular forces between layers

strong covalent bond

▲ Part of the structure of graphite. There are delocalised electrons between the layers.

H Fullerenes

Carbon can also exist as **fullerenes**. Fullerenes are a type of carbon made up of hexagonal rings of carbon atoms.

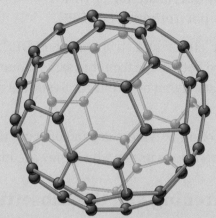

▲ One common fullerene is **Buckminsterfullerene**. It consists of molecules made up of 60 carbon atoms, arranged to form a hollow sphere.

Fullerenes are hollow. The space inside is big enough for atoms and small molecules to fit in. This means that fullerenes can be used to deliver drugs to specific places in the body. Fullerenes are also useful as lubricants, and as catalysts.

Fullerene molecules join up to make **nanotubes**. Nanotubes are used to reinforce materials, for example, in graphite tennis racquets.

Nanoscience

Nanoparticles are particles that are made up of a few hundred atoms. They have diameters of between 1 nanometre (nm) and 100 nm. **Nanoscience** is the study of nanoparticles.

Nanoparticles have different properties to those of the same substance in normal-sized pieces. Their surface area is high compared to their volume.

The special properties of nanoparticles may lead to new developments, including:

- new computers
- new catalysts
- new waterproof coatings
- stronger and lighter construction materials
- new cosmetics such as sun-protection creams and deodorants.

Exam tip

You need to know some applications of nanoscience, but you do not need to remember specific examples or properties.

Questions

1. Describe the bonding in diamond.

2. Explain why diamond and graphite have different properties.

3. List **five** ways in which nanoparticles may be used in the future.

Revision objectives

- ✓ explain how the properties of polymers are linked to what they are made from, and the conditions under which they are made
- ✓ explain how the uses of polymers are linked to their structures

Student book references

2.11 Explaining polymer properties

Specification key

✓ C2.2.5

Key words

high-density poly(ethene), low-density poly(ethene), thermosoftening polymer, thermosetting polymer

Exam tip AQA

Practise explaining the difference in properties between thermosoftening and thermosetting polymers.

Question

1 Make a table to summarise the differences in properties and structure between thermosoftening and thermosetting polymers.

Two types of poly(ethene)

There are two types of poly(ethene):
- **high-density poly(ethene)**, or HDPE
- **low-density poly(ethene)**, or LDPE.

HDPE is denser, stiffer, and stronger than LDPE. The two types of polythene are both made from the same monomer, ethene. But they are made using different catalysts and under different conditions.

The properties of all polymers depend on what they are made from and the conditions under which they are made.

Thermosoftening and thermosetting polymers

You can divide polymers into two groups:
- **Thermosoftening polymers** – soften easily when warmed, and can easily be moulded into new shapes. They can be recycled.
- **Thermosetting polymers** – do not melt when they are heated. They cannot be recycled.

The structures of the two types of polymer explain their properties. Thermosoftening polymers consist of individual, tangled polymer chains.

H The forces of attraction between the separate chains are weak.

weak forces between the separate polymer chains

▲ Polymer chains in a thermosoftening polymer.

Thermosetting polymers consists of polymer chains with cross-links between them. The cross-links prevent them from melting.

H The cross-links are strong intermolecular bonds.

chains held together by strong bonds

▲ Polymer chains in a thermosetting polymer.

1 Highlight the correct word or phrase in each pair of **bold** words in the sentences that follow.

> Methane consists of simple molecules. It has a **high/low** melting point. It **does/does not** conduct electricity because its molecules **have/do not have** an overall electric charge.

2 Highlight the statements below that are true. Then write corrected versions of the statements that are false.

   a The forces between the oppositely charged ions in ionic compounds are weak.

   b Ionic compounds have high boiling points.

   c Ionic compounds conduct electricity when solid.

   d Ionic compounds do not conduct electricity when dissolved in water.

   e When ionic compounds conduct electricity, the ions carry the current.

3 Choose words from the box below to fill in the gaps in the sentences that follow. The words in the box may be used once, more than once, or not at all.

softer	stiff
molecules	atoms
bendy	more
harder	less

The layers of _____ in metals can slide over each other easily. This makes metals _____.
Alloys are _____ than pure metals because the different-sized atoms in the structure make it _____ easy for the layers to slide over each other.

4 **a** Give the name of **one** shape-memory alloy.

   b Give **one** use of this shape-memory alloy.

5 Use the data in the table to answer the questions below it.

Substance	Does the solid conduct?	Does the liquid conduct?	Melting point (°C)	Boiling point (°C)
J	No	Yes	801	1413
K	No	No	−182	−162
L	Yes	Yes	1083	2595
M	No	No	> 3550	4837
N	No	Yes	2852	3600

   a Identify **two** ionic compounds in the table.
   b Which letter represents copper?
   c Which letter represents diamond?
   d Predict the hardest substance in the table.
   e Identify the **two** substances with covalent bonds.
   f Predict the substance that can easily be bent.

6 Explain why:
   a thermosoftening polymers can be recycled
   b thermosetting polymers cannot be recycled
   c low-density poly(ethene) and high-density poly(ethene) have different properties.

7 The diagrams below show the structures of graphite and diamond. Use the diagrams to explain why diamond is hard and why graphite is soft and slippery.

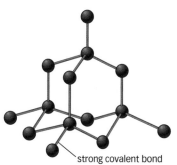

strong covalent bond

▲ Diamond.

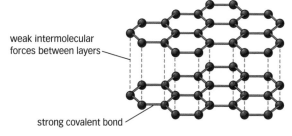

weak intermolecular forces between layers

strong covalent bond

▲ Graphite.

8 a Complete the sentences below.
  Fullerenes are a form of the element _____.
  The structure of fullerenes is based on _____ rings of atoms of this element.

b List **four** uses of fullerenes.

9 Write an **M** next to the sentences below that are true of metals only. Write a **G** next to the sentences that are true of graphite only. Write a **B** next to the sentences that are true of both metals and graphite.

a The structure includes delocalised electrons. ___

b There are strong covalent bonds between the atoms. ___

c The structure is arranged in layers. ___

d The structure includes positive ions. ___

e The layers can slide over each other. ___

10 The table shows the melting and boiling points of two substances whose atoms are joined together by covalent bonds.
  Use ideas about bonding and intermolecular forces to explain the differences in melting and boiling points.

Substance	Melting point (°C)	Boiling point (°C)
silicon dioxide	1610	2230
nitrogen dioxide	−11	21

11 The table gives data about two different forms of carbon: graphite and diamond.
  Use ideas about delocalised electrons to explain the difference shown.

Substance	Does it conduct electricity?
diamond	no
graphite	yes

12 Use ideas about intermolecular forces to explain why thermosoftening polymers melt at low temperatures, and why thermosetting polymers do not melt when they are heated.

Examination questions
Structure, properties, and uses

1 a The table below gives data about the properties of low-density poly(ethene) (LDPE) and high-density poly(ethene) (HDPE).

	LDPE	HDPE
density (g/cm^3)	0.92	0.95
strength (MPa)	12	31
transparency	good transparency	less transparent
relative flexibility	flexible	stiff

Draw a (ring) around the correct answer in each box to complete the sentence.

HDPE is more suitable for making garden furniture than LDPE

because HDPE is [weaker / stronger / less transparent] and [stiffer. / more flexible. / less dense.]

(2 marks)

b The diagram shows part of a poly(ethene) molecule.

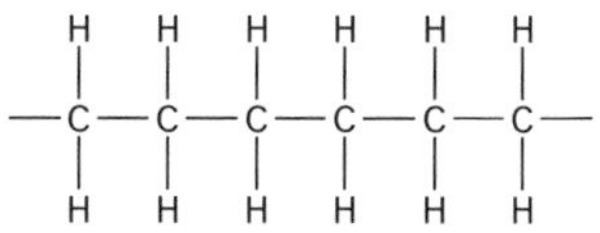

part of a poly(ethene) molecule

Draw a (ring) around the correct answer in each box to complete each sentence.

i Each hydrogen atom is joined to [1 / 2 / 4] other atom(s) .

(1 mark)

ii The bonds between the atoms in the molecule are [covalent. / metallic. / ionic.]

(1 mark)

iii A piece of LDPE consists of many individual polymer chains.

This means it is a [thermosetting / thermosoftening / cross-linked] polymer.

(1 mark)

(Total marks: 5)

2 Read this article, then answer the questions below it.

> Carbon nanotubes are a form of carbon. The carbon atoms are joined together in tiny tubes. The properties of carbon nanotubes are very different to the properties of other forms of carbon, such as graphite.
>
> Carbon nanotubes are very strong when subjected to pulling forces. They are also extremely stiff. They conduct heat well.
>
> A group of scientists have done experiments and found out that carbon nanotubes can enter human cells kept in test tubes. This makes the cells die. Studies on mice and rats suggest that inhaling carbon nanotubes over weeks or months may cause lung problems.

a Draw a ring around the correct answer below.

The diameter of a typical nanotube is

1–10 nm   **1–10 mm**   **1–10 cm**   **1–10 m**

(1 mark)

b Carbon nanotubes can be used to reinforce the materials used to make wind turbines.

Give **two** properties of carbon nanotubes that make them suitable for this purpose.

1 ....

2 ....

(2 marks)

c **i** Identify **two** possible health risks to humans who make or use carbon nanotubes.

1 ....

2 ....

(2 marks)

ii Suggest why we cannot be certain that each of these health problems will occur when a person is exposed to carbon nanoparticles.

....

....

(1 mark)
(Total marks: 6)

3 Rhodium is a metal.

a Small amounts of rhodium are added to platinum metal.

The resulting alloy is very hard.

Explain why the platinum-rhodium alloy is harder than pure platinum.

Your answer should include details of the following:
- How the atoms are arranged in pure platinum and in the platinum–rhodium alloy.
- Why pure platinum is relatively soft.
- Why the platinum–rhodium alloy is harder.

....

....

....

....

(4 marks)

H **b** Rhodium is used to coat electrical contacts.

Explain why rhodium is a good conductor of electricity.

....

....

(2 marks)

(Total marks: 6)

Revision objectives

- ✔ work out the mass number and atomic number of an atom
- ✔ explain what an isotope is
- ✔ calculate relative formula mass

Student book references

2.12 Atomic structure

2.13 Masses and moles

Specification key

- ✔ C2.3.1

Inside atoms

Atoms are made up of tiny **sub-atomic particles**. In the centre of an atom is its **nucleus**. This is made up of **protons** and **neutrons**. The nucleus is surrounded by **electrons**.

Name of particle	Relative mass	Relative charge
proton	1	+1
neutron	1	0
electron	very small	−1

Representing atoms

You can represent a sodium atom like this:

$$^{23}_{11}\text{Na}$$

- The **atomic number** is the number of protons in an atom. The atomic number of sodium is 11.
- The **mass number** is the total number of protons and neutrons in an atom. The atom of sodium above has 11 protons and 12 neutrons. Its mass number is $(11 + 12) = 23$.

Isotopes

Atoms of the same element can have different numbers of neutrons. This gives them different mass numbers. Atoms of an element that have different numbers of neutrons are called **isotopes**.

Oxygen has three naturally occurring isotopes. The table below shows the number of neutrons in one atom of each isotope.

Isotope	Number of protons	Number of neutrons	Mass number
^{16}O	8	8	16
^{17}O	8	9	17
^{18}O	8	10	18

H Relative atomic mass

The **relative atomic mass, A_r,** of an element compares the mass of atoms of the element with the mass of atoms of the ^{12}C isotope.

Relative atomic mass is an average value for the isotopes of an element. For example, copper has two isotopes:

- About 25% of copper atoms are of the ^{65}Cu isotope.
- About 75% of copper atoms are of the ^{63}Cu isotope.

The relative atomic mass of copper is 63.5. This is an average of the masses of the two isotopes, taking into account their relative amounts.

Relative formula mass

The formula of a substance tells you the number of each type of atom in the substance. The chemical dopamine is released by the brain when people fall in love. Its formula is $C_8H_{11}NO_2$. The formula tells you that one molecule of dopamine is made up of:

- 8 carbon atoms
- 11 hydrogen atoms
- 1 nitrogen atom
- 2 oxygen atoms.

The **relative formula mass (M_r)** of a substance is the sum of the relative atomic masses of the atoms in the numbers shown in the formula. You can work it out by adding together all the A_r values for the atoms in the formula. The relative formula mass of dopamine is

$$(12 \times 8) + (1 \times 11) + 14 + (16 \times 2) = 153.$$

Moles

The relative formula mass of a substance compares the mass of the substance to the mass of an atom of ^{12}C. You can also compare the masses in grams. The relative formula mass of a substance, in grams, is known as one **mole** of the substance. So the mass of one mole of carbon atoms is 12 g. The mass of one mole of dopamine molecules is 153 g.

Exam tip

Practise working out mass numbers and atomic numbers, and calculating relative formula masses.

Questions

1   Give the atomic numbers and mass numbers of these atoms:

$^{14}_{7}N$ $^{56}_{26}Fe$ $^{80}_{35}Br$

2   Explain the meaning of the word 'isotope'.

3   Calculate the relative formula mass of lead nitrate, $Pb(NO_3)_2$. Use data from the periodic table to help you.

Revision objectives

✓ calculate the percentage of an element in a compound

✓ **H** calculate empirical formulae, masses from equations, and percentage yields

Student book references

2.16 Using equations

2.17 Calculating yield

Specification key

✓ C2.3.3

Percentage by mass

To calculate the **percentage by mass** of an element in a compound, use the equation:

$$\frac{\text{percentage}}{\text{by mass}} = \frac{\text{relative mass of element in compound}}{\text{relative formula mass}} \times 100$$

> **Worked example**
>
> **Q** What is the percentage by mass of nitrogen in ammonium nitrate (NH_4OH)?
>
> **A** The relative formula mass, $M_r = 14 + (1 \times 4) + 16 + 1 = 35$
> The relative mass of nitrogen in the formula = 14
>
> Percentage of nitrogen in ammonium nitrate $= \dfrac{14}{35} \times 100 = 40\%$

H Empirical formulae

The **empirical formula** of a compound gives the relative number of the atoms of each element that are in the compound.

> **Worked example**
>
> **Q** A sample of a compound contains 0.20 g of hydrogen and 1.6 g of oxygen. What is its formula?
>
> **A**
>
> | | hydrogen | oxygen |
> |---|---|---|
> | mass of each element, in g | 0.20 | 1.6 |
> | A_r from periodic table | 1 | 16 |
> | mass divided by A_r | 0.2 | 0.1 |
> | simplest ratio | 2 | 1 |
>
> The formula of the compound is H_2O.

H Calculating masses of reactants and products from equations

In a chemical reaction, the total mass of reactants is equal to the total mass of products. This means you can calculate the masses of reactants and products from balanced symbol equations.

Worked example

Q 32 g of methane burns in air. What are the maximum masses of carbon dioxide and water that can be produced?

A $CH_4 + 2O_2 \rightarrow CO_2 + 2H_2O$

M_r of $CH_4 = 12 + (1 \times 4) = 16$

M_r of $CO_2 = 12 + (16 \times 2) = 44$

M_r of $H_2O = (1 \times 2) + 16 = 18$

Use ratios to work out the answer:

16 g of methane makes 44 g of carbon dioxide.

So 32 g of methane makes $(44 \times 2) = 88$ g of carbon dioxide and $(18 \times 2) = 36$ g of water.

Yield

In a chemical reaction, atoms cannot be gained or lost. But you don't always make the calculated amount of product. This may be because:

- you lose some of the product when separating it from the reaction mixture – in filtering, for example
- a reactant reacts in an unexpected way, for example, when burning in air, a substance might react with nitrogen as well as oxygen
- the reaction may be **reversible** – the products react to make the original reactants. For example:

    ammonium chloride $\rightleftharpoons$ ammonia + hydrogen chloride

The amount of product formed in a reaction is the **actual yield**. You can compare the actual yield to the **maximum theoretical yield**, calculated from the balanced symbol equation, to find the **percentage yield** of a reaction.

Exam tip

AQA

Practise finding relative atomic masses from different elements in the periodic table.

Worked example

Q Jasmine heated a piece of magnesium in air. She calculated a maximum theoretical yield for magnesium oxide of 8 g. But the actual yield was only 6 g. Calculate the percentage yield.

A percentage yield = $\dfrac{\text{actual yield}}{\text{maximum theoretical yield}} \times 100$

$= \dfrac{6\,g}{8\,g} \times 100 = 75\%$

Questions

1 Calculate the percentage of oxygen in magnesium oxide, MgO.

2 Explain why the maximum theoretical yield of a product is not always obtained in a chemical reaction.

3 **H** A sample of a compound contains 1.20 g of carbon and 0.4 g of hydrogen. What is its formula?

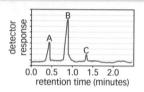

Key words

retention time, mass spectrometer, molecular ion peak

H A mass spectrometer produces a mass spectrograph like this for each compound that it analyses.

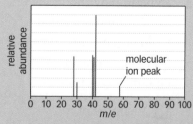

▲ The peak on the right is the molecular ion peak. The mass of the molecular ion gives the relative molecular mass of the compound.

Chemical analysis

You can use paper chromatography to separate artificial food colours.

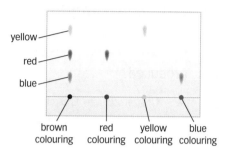

The chromatogram shows that ▶ the brown colouring is a mixture of blue, red, and yellow dyes.

Gas chromatography mass spectroscopy (GC-MS)

GC-MS is an instrumental method of analysis. Instrumental methods identify small samples. They are sensitive, accurate, and quick.

Gas chromatography separates mixtures. The diagram shows how it works.

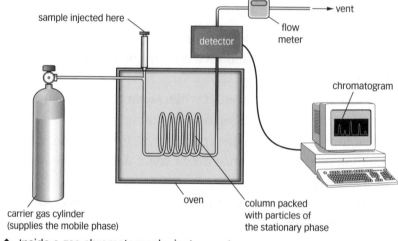

▲ Inside a gas chromatography instrument.

The gas chromatogram on the left (above) shows that:
- there are three peaks, so the mixture has three compounds
- compound A travelled most quickly, and left the column first. This means that compound A has the shortest **retention time**. Retention times help identify substances.
- peak B has the greatest area, so compound B is present in the mixture in the greatest amount.

The gas chromatography apparatus may be connected to a **mass spectrometer**. This identifies the separated compounds.

Questions

1 List **three** advantages of instrumental analysis methods.

2 Explain how gas chromatography apparatus separates the compounds of a mixture.

Working to Grade E

1   Draw lines to match each type of particle to its relative mass.

Type of particle
proton
neutron
electron

Relative mass
very small
1
1

2   Tick the boxes to show the benefits of instrumental methods of chemical analysis.
   a   They are quick.
   b   The instruments are expensive.
   c   They are accurate.
   d   They can identify tiny amounts of substances.
   e   People need to be highly trained to operate the instruments.

3   The statements below describe the steps in using GC-MS to separate and identify compounds. Write the letters of the steps in the best order.
   A   The different vapours in the mixture travel through the column, which is packed with a solid material, at different speeds.
   B   A liquid mixture of unknown compounds is injected into the gas chromatography instrument.
   C   The vapours leave the column at different times.
   D   A carrier gas mixes with the vapours.
   E   This separates the vapours in the mixture.
   F   The mass spectrometer identifies the substances as they leave the chromatography column.
   G   The liquid mixture is heated so that it becomes a mixture of vapours.

4   Highlight the statement or statements below that are true. Then write corrected versions of any that are false.
   a   Isotopes are atoms of the same element that have different numbers of neutrons.
   b   The relative formula mass of a substance, in grams, is called one squirrel of that substance.
   c   The theoretical yield of a substance in a chemical reaction is always the same as, or greater than, the actual yield.

Working to Grade C

5   Give the atomic numbers and mass numbers of the atoms listed below.
   a $^{40}_{18}Ar$ b $^{55}_{25}Mn$ c $^{65}_{30}Zn$
6   Calculate the masses of one mole of:
   a   paracetamol, $C_8H_9NO_2$
   b   aspirin, $C_9H_8O_4$

7   Calculate the percentage by mass of:
   a   potassium in potassium cyanide, KCN
   b   lithium in lithium carbonate, Li_2CO_3
   c   nitrogen in serotonin, $C_{10}H_{12}N_2O$

8   Use the gas chromatograph to answer the questions.

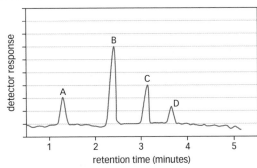

   a   How many compounds were in the mixture?
   b   Which substance has the longest retention time?
   c   Which substance travelled most quickly?

9   Give **three** reasons to explain why it is not always possible to obtain the maximum theoretical yield of a product in a chemical reaction.

Working to Grade A*

10   The diagram below is a simplified mass spectrum of a substance. What is the molecular mass of the substance?

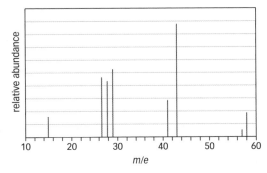

11   Write a definition for the relative atomic mass of an element, A_r.

12   Calculate the formulae of samples of compounds that contain:
   a   3.2 g of sulfur and 3.2 g of oxygen
   b   2.4 g of carbon and 0.4 g of hydrogen
   c   2.3 g of sodium, 1.4 g of nitrogen and 4.8 g of oxygen.

13   Calculate the maximum theoretical yield of carbon dioxide, if a student heats 10.0 g of calcium carbonate. The equation for the reaction is $CaCO_3 \rightarrow CaO + CO_2$

14   Miss Corner heats a known mass of potassium in chlorine. The maximum theoretical yield of potassium chloride is 1.5 g. The actual yield is 1.0 g. Calculate the percentage yield.

85

Examination questions
Atomic structure, analysis, and quantitative chemistry

1 a An atom of the element zirconium may be represented like this:

$$^{91}_{40}\text{Zr}$$

    i Give the mass number of this zirconium atom.

....

(1 mark)

    ii Give the number of protons in an atom of zirconium.

....

(1 mark)

b Two more atoms of zirconium are represented like this:

$$^{92}_{40}\text{Zr and }^{94}_{40}\text{Zr}$$

    i Give the number of neutrons in each of these zirconium atoms.

$^{92}_{40}\text{Zr}$ ....

$^{94}_{40}\text{Zr}$ ....

(2 marks)

    ii Draw a ring around the correct answer to complete the sentence below.

Different atoms of zirconium have different numbers of neutrons.

These atoms are called | molecules / ions / isotopes | of zirconium.

(1 mark)

(Total marks: 5)

2 Morphine is made in the body after the injection of the drug heroin.

a The formula of morphine is $C_{17}H_{19}NO_3$.

Calculate its relative formula mass, M_r.

....

....

(2 marks)

b The formula of heroin is $C_{21}H_{23}NO_5$.

Its relative formula mass, M_r, is 369.

    i What is the mass of one mole of heroin?

Include the unit in your answer.

....

(1 mark)

ii Calculate the percentage by mass of oxygen in heroin.

....

....

....

(2 marks)

c Morphine can be detected in the hair of heroin users.

The diagram shows part of a simplified gas chromatogram obtained by analysing the hair of a heroin user.

The heroin user had also taken the drug codeine.

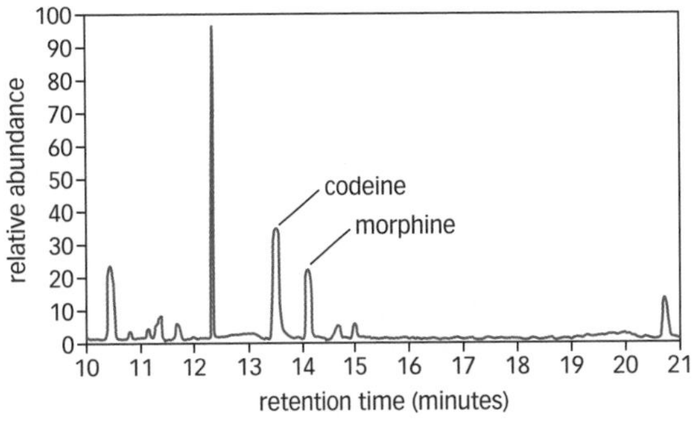

i Which travels more slowly through the column in the gas chromatography apparatus, morphine or codeine?

Give a reason for your decision.

....

....

(1 mark)

ii Draw a ring around the correct answer to describe what happens in the column of the gas chromatography apparatus.

The mixture is separated.

The relative molecular masses of the compounds in the mixture are measured.

Liquids are vapourised. *(1 mark)*

(Total marks: 7)

H **3** A chemist heated 1.2 g of magnesium in air.

She used the equation below, and data from the periodic table, to calculate the maximum theoretical yield of magnesium oxide as 2.0 g.

$$2Mg(s) + O_2(g) \rightarrow 2MgO(s)$$

She actually obtained only 1.5 g of magnesium oxide.

a Calculate the percentage yield of magnesium oxide.

....

....

(2 marks)

b Suggest why the actual yield of magnesium oxide was less than the maximum theoretical yield.

....

....

(1 mark)

(Total marks: 3)

Societal aspects of scientific evidence

Decisions about scientific issues are not usually based on evidence alone. Other factors, such as those relating to ethical, social, economic or environmental concerns, are also taken into account when evaluating the impacts of new developments.

Materials for devices to keep blood vessels open

Skill – Analysing the facts and making deductions

In this question you will be assessed on using good English, organising information clearly, and using specialist terms where appropriate.

Read the information below and answer the questions that follow.

A doctor may insert a stent into a blood vessel that has become narrow or blocked. The stent helps to keep the blood vessel open.

Some stents are made from stainless steel. Others are made from a shape-memory alloy called Nitinol. At body temperature, Nitinol stents change shape to match the shape of the blood vessels they are holding open. Stainless steel does not change shape in this way. Nitinol stents return to their original shape after being squashed. Stainless steel stents do not.

Research suggests that blood clots are less likely to form on Nitinol stents than on stainless steel ones.

Nitinol is an alloy of nickel and titanium. Some people are allergic to nickel, and it is possible that, in these people, new blockages might form in the blood vessel near the stent. If nickel compounds get into the blood stream, the risk of cancer may increase.

However, if the surface of a Nitinol stent is treated properly, a layer of titanium dioxide forms on its surface. This means that nickel is very unlikely to get into the bloodstream of a person with a stent.

Stainless steel is mainly iron. It contains small amounts of other metals. Scientists have shown that if people are allergic to these metals, new blockages might form in the blood vessel near the stent.

1   Use the information in the box to evaluate the use of stainless steel compared with Nitinol as a material for making stents.

This question includes the word *evaluate*. To answer it, you will need to write down some advantages and disadvantages of making stents from Nitinol, and some advantages and disadvantages of making stents from stainless steel. You will then need to compare the advantages and disadvantages, and write down which material is better for stents, and why.

Before you start writing your answer, you may find it helpful to organise your ideas in a table like the one below.

	Advantages	Disadvantages
Nitinol		
Stainless steel		

Once you've organised your ideas like this, you can decide which you think is the better material, and why.

Of course, there is no one 'correct answer' to this question. You can get full marks whatever your decision, provided you have stated the advantages and disadvantages clearly, and given reasons linked to these for your final decision.

2   Suggest one economic factor that a hospital might consider before deciding which of the two types of stent to offer its patients.

To answer this question, you will need to think of a factor that is related to money.

Skill – Evaluating bias

3   A student makes notes about four scientists who have done research about stents and the materials they are made from.

Anita Smith – 22 years old, funding herself through university.
Professor Nadeem Hanif – university scientist with 25 years experience. Government money funds his research.
Dr Bernard Anning – scientist with 30 years experience. Works for company that makes Nitinol.
Dr Rachel Hooper – hospital doctor who has inserted many stainless steel and Nitinol stents into patients over the past two years, and has collected data on their health after having the stents. Research funded by company that makes stainless steel stents.

a   The student says that she has greater trust in the research findings of Anita Smith than of Rachel Hooper. Suggest **one** reason for this opinion.

b   Using only the information in the box above, name the scientist whose evidence you think other scientists should pay most attention to. Give a reason for your decision.

In answering parts a and b you will need to consider whether the evidence from any of the scientists might be biased, and if so, why. You also need to think about the status of the researchers – for example, who is more experienced? For part b there is no one correct answer – the examiner is more interested in the reasons you give to back up your decision.

AQA Upgrade

Answering a question using formulae

QUESTION

The compound strontium chloride is used in red fireworks. It is also added to seawater aquaria, where it is used by tiny sea creatures to make their skeletons.

1 This diagram shows the electronic structure of a chlorine atom. Draw the structure of the chloride ion, Cl⁻. *(2 marks)*

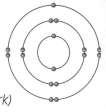

2 The formula of the strontium ion is Sr^{2+}. Give the formula of strontium chloride. *(1 mark)*

3 Explain why solid strontium chloride does not conduct electricity, but a solution of strontium chloride does conduct electricity. *(4 marks)*

1

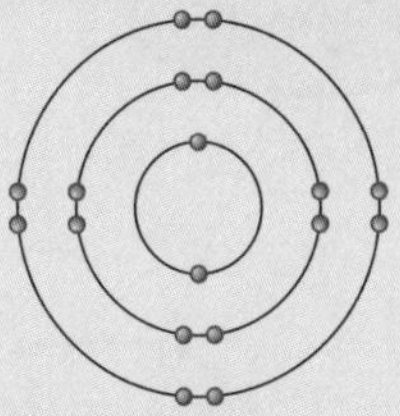

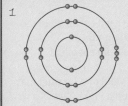

2 $SrCl_2$

3 In solid strontium chloride, there are strong electrostatic forces in all directions between the oppositely charged ions. The ions are not free to move to carry an electric current. In a solution, the charged particles are free to move and carry the current. So a solution of strontium chloride does conduct electricity.

B–A*

Examiner: This answer gains seven marks out of seven. All parts are answered correctly.

The explanation given in answer to question 3 is accurate and detailed.

1

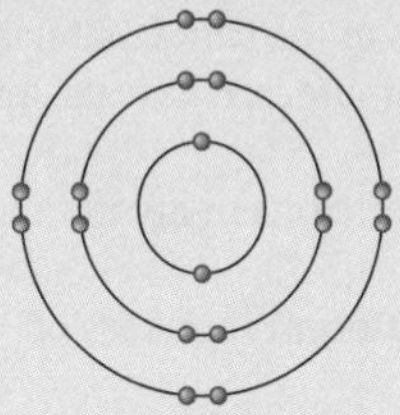

2 Sr_2Cl

3 The solid doesn't conduct because it has no ions that are free to move. The solution has moving ions so it conducts.

D–C

Examiner: This answer gains four marks out of seven.

Question 1 is answered correctly, and gains two marks.

Question 2 is incorrect, and gains zero marks. The candidate has not checked that the total number of positive charges on the ions shown in the formula is equal to the total number of negative charges on the ions shown in the formula.

The student gains two marks out of four for question 3, since each part of it is answered correctly, but in insufficient detail.

1

2 $SrCl$

3 The solid does not conduct because it is all held together by strong covalent bonds, but the solution does because the ions are free to move.

G–E

Examiner: This answer gains just one mark out of seven.

The structure given for question 1 is incorrect, since the highest occupied energy level contains nine electrons, instead of eight.

The formula for question 2 is also incorrect – two Cl⁻ ions are needed to balance the two positive charges on the Sr^{2+} ion.

One mark is awarded for the second half of the sentence that answers question 3. The first part of the answer is incorrect – solid strontium chloride is held together by ionic bonds, not covalent ones.

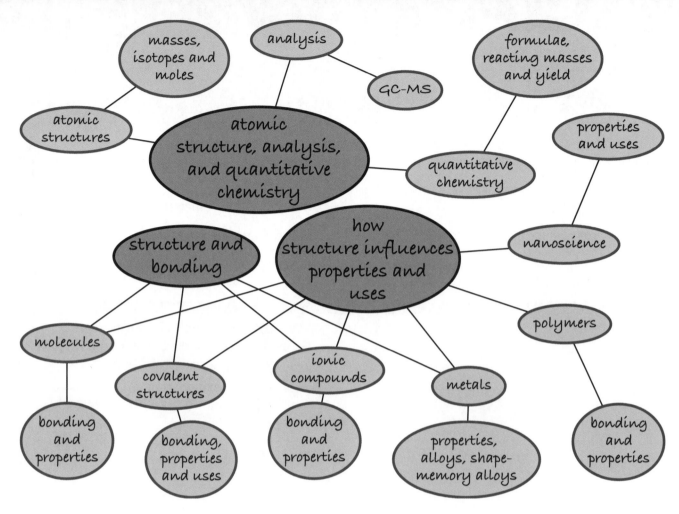

Revision checklist

- Chemical bonding involves transferring or sharing electrons in the highest occupied energy levels of atoms to achieve the stable electronic structure of a noble gas.
- Ions are formed when electrons transfer from one atom to another.
- Ionic compounds are held together by strong electrostatic forces of attraction. This gives them high melting and boiling points. They conduct electricity when molten or in solution.
- Strong covalent bonds form when atoms share electrons. Some covalently bonded substances consist of simple molecules. Others have giant structures of atoms.
- Substances with giant covalent structures include forms of carbon, such as graphite and diamond, and silicon dioxide. They have high melting points.
- Metals have giant structures of atoms arranged in layers in a regular pattern. The layers slide over each other, so metals are bendy.
- Metals have delocalised electrons in their structures, so they can conduct heat and electricity.
- Most alloys are made of two or more metals. Different sized atoms distort the layers, so alloys are harder than pure metals.
- Shape-memory alloys such as Nitinol return to their original shape after being deformed.

- Nanoparticles are structures of between 1 and 100 nm in size. They have different properties to the same material in bulk. They may be used in new computers, catalysts, and cosmetics.
- The total number of protons and neutrons in an atom is its mass number.
- Atoms of the same element with different numbers of neutrons are isotopes.
- The relative atomic mass of an element compares the mass of atoms of the element with the ^{12}C isotope.
- The relative formula mass (M_r) of a compound is the sum of the relative atomic masses of the atoms in the numbers shown in the formula.
- Gas chromatography linked to mass spectroscopy (GC-MS) is an instrumental method of analysis. It is accurate, sensitive, and quick.
- Information from formulae can be used to calculate the percentage of an element in a compound.
- Reacting masses can be calculated from chemical equations.
- In a chemical reaction, the actual yield of a product may be less than the maximum theoretical yield. The percentage yield can be calculated from these values.

Reaction rates

The **rate of reaction** is a measure of how quickly a reaction happens. To work out the rate of a reaction, you need to do experiments to find out how quickly products are made or reactants are used up.

Following reactions

This equation shows how magnesium reacts with hydrochloric acid.

$$\text{magnesium} + \frac{\text{hydrochloric}}{\text{acid}} \rightarrow \frac{\text{magnesium}}{\text{chloride}} + \text{hydrogen}$$

$$\text{Mg} + 2\text{HCl} \rightarrow \text{MgCl}_2 + \text{H}_2$$

You can follow the reaction by measuring the volume of hydrogen made as the reaction happens.

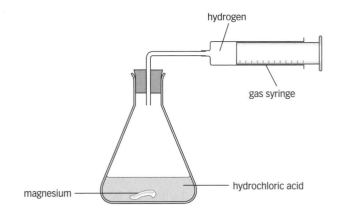

The graph on the right shows the volume of hydrogen made during the reaction.
- At first, the gradient is steep, showing that the reaction is fast.
- The gradient gets less steep over time, showing that the reaction slows down.

Calculating rates

You can use the equation below to calculate the rate of the reaction of magnesium with hydrochloric acid.

$$\text{rate of reaction} = \frac{\text{amount of product made}}{\text{time}}$$

From the graph, the rate of the reaction for the first minute

$$= \frac{25\,\text{cm}^3}{1\,\text{min}} = 25\,\text{cm}^3/\text{min}$$

Over the first three minutes, the average reaction rate

$$= \frac{40\,\text{cm}^3}{3\,\text{min}} = = 13.3\,\text{cm}^3/\text{min}$$

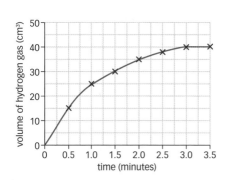

If you have data for the amount of reactant, you can use the equation below to calculate reaction rates:

$$\text{rate of reaction} = \frac{\text{amount of reactant used}}{\text{time}}$$

Collisions and activation energy

Reactions can only happen when reactant particles **collide**, or hit each other. The colliding particles must have enough energy to react.

The minimum amount of energy that particles need in order to react is the **activation energy**.

Temperature and reaction rate

Changing the reaction conditions changes the rate of a reaction.

Increasing the temperature increases the rate of a reaction. This is because higher temperatures make particles in a reacting mixture move faster. This makes their collisions:

- more frequent
- more energetic.

The more energetic a collision, the greater the likelihood of it leading to a reaction, and so being a **successful collision**.

Sodium thiosulfate reacts with hydrochloric acid according to the equation below. The sulfur formed in the reaction is a solid. It obscures a cross drawn on a piece of paper placed beneath the reaction flask.

sodium thiosulfate + hydrochloric acid → sodium chloride + water + sulfur dioxide + sulfur

The graph below shows the link between temperature and reaction time for the reaction. The shorter the reaction time, the faster the rate of reaction.

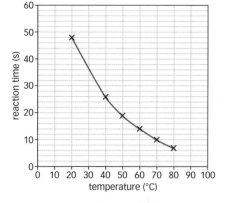

Pressure

You can speed up reactions involving gases by increasing the pressure.

Increasing the pressure of a mixture of gases makes the particles more crowded. This means they collide more frequently. Increasing the frequency of collisions increases the rate of the reaction.

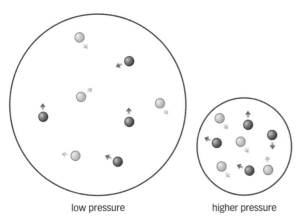

low pressure   higher pressure

▲ Gas particles collide more frequently at higher pressures.

Concentration

You can speed up a reaction involving a solution by increasing its concentration.

The more concentrated a solution, the greater the number of solute particles dissolved in a certain volume of the solution, and the more crowded the particles. Increased crowding leads to more frequent collisions, and a faster reaction.

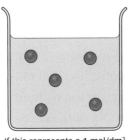

If this represents a 1 mol/dm³ solution of acid ...   then this represents a 2 mol/dm³ solution of the same acid.

▲ There are double the number of acid particles in the same volume of water. Water particles are not shown on the diagrams.

The graph opposite shows the link between reaction time and acid concentration for the reaction of hydrochloric acid with magnesium ribbon. Remember – the shorter the reaction time, the faster the reaction.

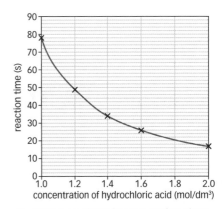

▲ Graph to show how reaction time varies with concentration.

Revision objectives

✔ describe and explain the effects of changing concentration, pressure and surface area on rates of reaction
✔ explain what catalysts do and why they are important in industry

Student book references

2.20 Speeding up reactions – concentration
2.21 Speeding up reactions – surface area
2.22 Speeding up reactions – catalysts

Specification key

✔ C2.4.1 d – h

Surface area

A reaction involving a powder happens faster than a reaction involving a lump of the same reactant. A powder has a bigger **surface area** than a lump of the same mass. This is because particles that were inside the lump become exposed on the surface when it is crushed.

The bigger the surface area, the greater the frequency of collisions, and the faster the reaction.

The equation below summarises the reaction of hydrochloric acid with calcium carbonate.

$$\text{hydrochloric acid} + \text{calcium carbonate} \rightarrow \text{calcium chloride} + \text{water} + \text{carbon dioxide}$$

You can use the apparatus on the left to follow the reaction by measuring the decrease in mass as carbon dioxide gas is made and given off.

The graph shows that the reaction is faster when powdered calcium carbonate is used.

Catalysts

You can increase the rate of many reactions by adding a **catalyst** to the reaction mixture. A catalyst increases the rate of a reaction without being used up in the reaction.

For example, hydrogen peroxide in solution breaks down very slowly to form water and oxygen:

$$\text{hydrogen peroxide} \rightarrow \text{water} + \text{oxygen}$$
$$2H_2O_2 \rightarrow 2H_2O + O_2$$

Adding powdered manganese(IV) oxide **catalyses**, or speeds up, the reaction, making the hydrogen peroxide decompose more quickly.

Catalysts are important in the chemical industry. They make chemical reactions fast enough to be profitable, and may reduce energy costs. Iron catalyses the reaction of hydrogen with nitrogen to make ammonia. Ammonia makes fertilisers and explosives.

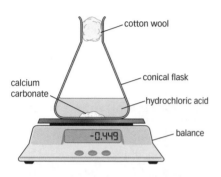

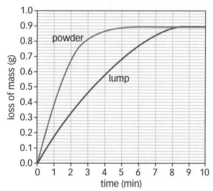

▲ The loss in mass during the reaction of calcium carbonate with hydrochloric acid.

Questions

1   List **four** factors that affect reaction rate.

2   Use ideas about collisions to explain why increasing the surface area of a solid reactant increases reaction rate.

3   Explain what catalysts do, and why they are important in industry.

Questions
Rates of reaction

Working to Grade E

1 Tick the boxes to show which of the following changes increase the rate of reaction.
   a Increasing the temperature
   b Diluting any solutions present
   c Increasing the gas pressure
   d Decreasing the surface area of any solids

2 What is the activation energy of a reaction? Tick the best definition in the table below.

Definition	Tick (✓)
The temperature change for the reaction.	
The energy change for the reaction.	
The maximum energy the particles must have to react.	
The minimum energy the particles must have to react.	

3 Write a definition of the word 'catalyst'.

4 A student adds hydrochloric acid to calcium carbonate (marble chips) in the apparatus below.

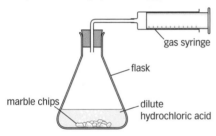

gas syringe

flask

marble chips

dilute hydrochloric acid

One of the products of the reaction is carbon dioxide gas. Every minute, the student records the volume of carbon dioxide gas in the gas syringe. The graph below shows the results.

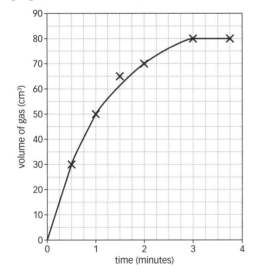

a Which description best describes how the volume of gas changes with time? Tick **one** description.

Description	
The volume of gas increases quickly at first, and then more slowly.	
The volume of gas increases quickly.	
The volume of gas increases slowly at first, and then more quickly.	
The volume of gas increases quickly.	

b Use data from the graph and the equation in the box below to calculate:
   i the rate of reaction in the first minute.
   ii the rate of reaction in the second minute.

$$\text{Rate of reaction} = \frac{\text{volume of product formed}}{\text{time}}$$

c Add a line to the graph above to show how the volume of gas would change with time if the student repeated the investigation at a higher temperature.

5 Use ideas about particles to explain why
   a increasing the temperature of a reaction increases its rate.
   b increasing the concentration of reactants in solution increases the rate of reaction.

6 A chemical company makes ammonia. Suggest why it uses a catalyst to speed up the reaction.

7 A student sets up the apparatus below.

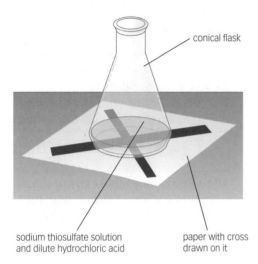

conical flask

sodium thiosulfate solution and dilute hydrochloric acid

paper with cross drawn on it

He measures the time taken for the cross below the flask to disappear. He repeats the investigation at a total of five temperatures. This table summarises his results.

Temperature (°C)	Time for cross to disappear (s)
20	400
30	200
40	100
50	47
60	26
70	60

a Identify the independent variable.
b Identify **three** control variables.
c Identify the range of the dependent variable.
d Plot the results on a graph and draw a line of best fit.
e Identify the anomalous result.
f Describe the pattern shown on the graph.

8 A student monitors the reaction of big lumps of calcium carbonate with dilute hydrochloric acid. Every 30 seconds, she measures the total volume of gas that has been formed.
She repeats the procedure with smaller pieces of calcium carbonate, and then with calcium carbonate powder. She keeps all the other reaction conditions the same.
This graph summarises the student's data.

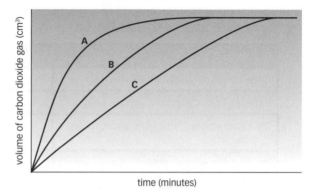

a Which curve on the graph represents the reaction with the biggest pieces of calcium carbonate?
b Explain why the total volume of carbon dioxide made is the same each time.

1 A student tested the hypothesis that there is a link between temperature and the rate of a reaction.

He added magnesium ribbon to dilute hydrochloric acid.

He measured the time for the magnesium ribbon to completely react, and disappear.

He repeated the investigation at a total of five different temperatures.

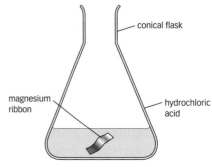

a List **two** variables that must be controlled in order to obtain valid results.

1 ....

2 ....

(2 marks)

b The student plotted his results on the graph below. Use the graph to answer the questions that follow.

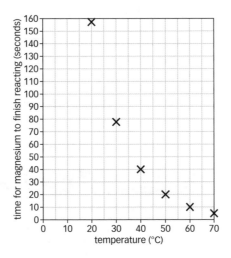

i Draw a line of best fit on the graph. *(1 mark)*

ii Use the graph to predict the time for the magnesium to disappear at 45 °C.

....

(1 mark)

iii Describe the pattern shown by the graph.

....

....

(2 marks)

iv Use ideas about particles to explain the link between temperature and reaction rate.

....

....

(2 marks)

(Total marks: 8)

Revision objectives

- explain what exothermic and endothermic reactions are, and give examples of each
- recall that reactions that are endothermic in one direction are exothermic in the other

Student book references

2.23 Energy and chemical reactions

2.24 Energy in, energy out

Specification key

✔ C2.5.1

Energy transfer

When chemical reactions happen, energy is transferred to or from the surroundings. The energy may be transferred as heat, light, sound or movement.

Exothermic reactions

An **exothermic reaction** is one that transfers energy to the surroundings. Examples of exothermic reactions include
- combustion reactions
- many oxidation reactions
- neutralisation reactions.

Combustion (burning) reactions transfer energy to the surroundings, mainly as heat and light. The energy transferred in combustion reactions is useful for cooking, heating homes, and generating electricity.

The neutralisation reaction of dilute hydrochloric acid with sodium hydroxide solution is exothermic. When you mix the two solutions, the temperature increases as the reaction takes place. The temperature then decreases slowly to room temperature as energy is transferred as heat from the reaction mixture to the surroundings.

Substance	Temperature (°C)
sodium hydroxide solution, before mixing	20
hydrochloric acid, before mixing	20
reaction mixture, immediately after mixing	49
reaction mixture, 1 hour after mixing	20

Using exothermic reactions

Exothermic reactions are used in items such as:
- Hand warmers – when you activate the hand warmer, there is an exothermic reaction. This transfers heat energy to your hands.
- Self-heating coffee cans – these have two compartments. One contains cold coffee. The other contains substances that react together in an exothermic reaction. The reaction transfers heat to the coffee.

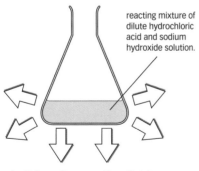

reacting mixture of dilute hydrochloric acid and sodium hydroxide solution.

▲ When the reaction finishes, energy is transferred to the surroundings.

Endothermic reactions

An **endothermic reaction** takes in energy from the surroundings.

Thermal decomposition reactions are endothermic. For example:

copper carbonate → copper oxide + carbon dioxide

$$CuCO_3 \rightarrow CuO + CO_2$$

Using endothermic reactions

Some sports injury packs are based on an endothermic reaction. When activated, there is a chemical reaction between the reactants in the pack. The temperature of the reacting mixture decreases, cooling the injury.

As the pack takes in heat from the injured muscle, the temperature of the pack gradually increases to the temperature of the surroundings.

Reversible reactions

If a reversible reaction is exothermic in one direction, it is endothermic in the opposite direction. The same amount of energy is transferred in each direction.

Blue copper sulfate crystals contain water – they are **hydrated**. Heating hydrated copper sulfate crystals forms white **anhydrous** copper sulfate powder, which contains no water. The process is endothermic.

Adding water to anhydrous copper sulfate powder is an exothermic process. Heat energy is transferred to the surroundings.

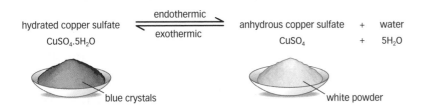

hydrated copper sulfate $\xrightleftharpoons[\text{exothermic}]{\text{endothermic}}$ anhydrous copper sulfate + water

$CuSO_4.5H_2O$ $CuSO_4$ + $5H_2O$

blue crystals   white powder

Questions

1   What is an exothermic reaction?

2   Give an example of **one** way in which endothermic reactions can be useful.

3   Explain why, in an exothermic reaction, the temperature first increases, and then decreases back to the temperature of the surroundings.

1 Choose words from the box below to fill in the gaps in the sentences that follow. The words in the box may be used once, more than once, or not at all.

> always never sometimes exothermic
> endothermic

In some chemical reactions, energy is transferred to the surroundings. These are _____ reactions. Some reactions take in energy from the surroundings. These are _____ reactions. Energy is _____ transferred to or from the surroundings in chemical reactions.

2 Draw lines to match each type of reaction to one or more uses.

Type of reaction
endothermic
exothermic

Use
hand warmers
sports injury packs
self-heating cans for coffee

3 Tick **one** box to show which type of reaction below is endothermic.
a Combustion ☐
b Oxidation ☐
c Neutralisation ☐
d Thermal decomposition ☐

4 A student neutralised an alkali with three different acids. She measured the temperature change in each neutralisation reaction.

Acid	Temperature change (°C)			
	Test 1	Test 2	Test 3	Mean
hydrochloric acid	25	26	27	26
nitric acid	20	5	20	20
ethanoic acid	14	9	10	

a Calculate the missing mean.
b Identify the anomalous result.
c Why did the student do three tests for each acid? Tick (✓) **two** answers.

Reason	Tick (✓)
To improve the resolution of the data	
To spot any anomalous data	
To spot any unreliable results	
To improve the accuracy of the data	

d The student wanted to make valid comparisons between the two acids. Which variables should she control? Tick the correct answers.

Variable	Tick (✓)
The volume of the acid	
The temperature of the acid	
The type of acid	
The concentration of the acid	

5 Put a tick next to each of the reactions below that is likely to be exothermic.

	Reaction	Tick (√)
(a)	magnesium + oxygen → magnesium oxide	
(b)	copper → copper + carbon carbonate oxide dioxide	
(c)	methane + oxygen → carbon dioxide + water	
(d)	hydrogen + oxygen → water	
(e)	lithium → lithium + nitrogen + oxygen nitrate oxide dioxide	

6 A student mixed five pairs of solutions. On mixing, reactions occurred. The student measured the temperatures of the solutions before and immediately after mixing. His results are in the table.

Reaction	Temperature before reaction (°C)	Temperature immediately after mixing (°C)
A	20	54
B	20	71
C	20	5
D	20	82
E	20	14

An hour later, the student measured the temperatures of the reaction mixtures again. The temperature of each mixture was 20 °C, the same as the temperature of the room.
a Give the letters of the reaction mixtures that transfer energy to the surroundings as they return to room temperature.
b Give the letters of the reaction mixtures that take in energy as they return to room temperature.
c Give the letters of reactions that are endothermic.
d Give the letters of the reactions that are exothermic.

7 Annotate the equation below to show the direction in which the reaction is exothermic and the direction in which the reaction is endothermic.

hydrated copper sulfate $\rightleftharpoons$ anhydrous copper sulfate + water

Examination questions
Exothermic and endothermic reactions

1 A student neutralised an acid with an alkali.

She recorded the temperatures of the solutions before and after the reaction.

Her results are in this table.

temperature of acid before reaction (°C)	19
temperature of alkali before reaction (°C)	21
maximum temperature reached after reaction (°C)	69

a Explain how the results show that the neutralisation reaction is exothermic.

....

....

(1 mark)

b Which other types of reaction are usually exothermic?

Tick (✓) **two** boxes.

Type of reaction	Tick (✓)
thermal decomposition	
oxidation	
combustion	

(1 mark)

c Give an example of **one** way in which exothermic reactions are useful.

....

(1 mark)

d **i** Describe what happens to the temperature of a sports injury pack immediately after it is activated.

....

(1 mark)

ii Explain why the temperature of the sports injury pack gradually returns to that of the surroundings.

....

....

(1 mark)

(Total marks: 5)

Revision objectives

- ✔ explain what makes solutions acidic or alkaline
- ✔ explain what happens in neutralisation reactions

Student book references

2.25 Acids and bases

Specification key

- ✔ C2.6.1 a
- ✔ C2.6.2 a, c – e

State symbols

In symbol equations, **state symbols** give the states of the reactants and products:

- (s) means solid
- (l) means liquid
- (g) means gas
- (aq) means aqueous, or dissolved in water.

Bases, alkalis, and acids

Metal oxides and metal hydroxides are **bases**. Bases neutralise acids. Copper oxide, zinc oxide, and sodium hydroxide are examples of bases.

Some metal hydroxides are soluble in water. These hydroxides are **alkalis**. Sodium hydroxide and potassium hydroxide are examples of alkalis.

Hydroxide ions, OH^-(aq) make solutions alkaline.

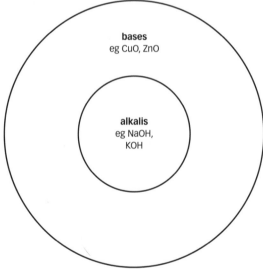

▲ All metal oxides and hydroxides are bases. Soluble hydroxides are called alkalis.

Ammonia

Ammonia, NH_3, dissolves in water to make an alkaline solution.

$$\text{ammonia} + \text{water} \rightarrow \text{ammonium hydroxide}$$
$$NH_3(g) \ + H_2O(l) \rightarrow \qquad NH_4OH(aq)$$

In solution, the ammonium ions (NH_4^+) and the hydroxide ions (OH^-) are separated and surrounded by water molecules. The dissolved OH^- ions make the solution alkaline.

Compounds containing ammonium ions are important fertilisers.

pH scale

The **pH scale** is a measure of the acidity or alkalinity of a solution. On the pH scale:

- a solution of pH 7 is neutral
- a solution with a pH of less than 7 is acidic
- a solution with a pH of more than 7 is alkaline.

Key words

state symbol, base, alkali, hydroxide ion, pH scale, neutralisation reaction

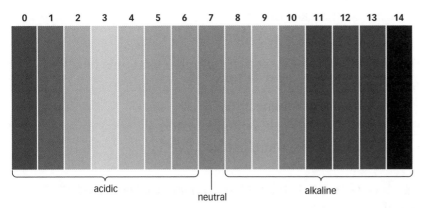

0 1 2 3 4 5 6 7 8 9 10 11 12 13 14

acidic   neutral   alkaline

▲ The pH scale. Hydroxide ions (OH⁻) make solutions alkaline and hydrogen ions (H⁺) make solutions acidic. Universal indicator is usually red, orange, or yellow in acidic solutions, green in neutral solutions, and blue or purple in alkaline solutions.

Neutralisation reactions

You can use sodium hydroxide solution to neutralise hydrochloric acid. The products are sodium chloride and water:

$$\text{hydrochloric acid} + \text{sodium hydroxide} \rightarrow \text{sodium chloride} + \text{water}$$

$$HCl(aq) + NaOH(aq) \rightarrow NaCl(aq) + H_2O(l)$$

This is an example of a **neutralisation reaction**. In the reaction, OH⁻ ions from the sodium hydroxide react with H⁺ ions from the acid to form water. You can represent neutralisation reactions like this:

$$H^+(aq) + OH^-(aq) \rightarrow H_2O(l)$$

Exam tip

Remember, an alkali is a soluble base. So all alkalis are bases, but not all bases are alkalis.

Questions

1   Name **five** bases and **two** alkalis.

2   Explain why ammonia dissolves in water to make an alkaline solution.

3   Write a word equation for the neutralisation reaction of hydrochloric acid with potassium hydroxide.

4   **H** Write a symbol equation for the reaction of hydrochloric acid with potassium hydroxide. Include state symbols.

Revision objectives

- ✓ explain what a salt is
- ✓ describe how to make soluble salts from acids and metals
- ✓ describe how to make soluble salts from acids and metal oxides
- ✓ describe how to make soluble salts from acids and alkalis

Student book references

2.26 Making soluble salts – 1

2.27 Making soluble salts – 2

Specification key

- ✓ C2.6.1 b – c
- ✓ C2.6.2 b

Salts

A **salt** is a compound that contains metal or ammonium ions. Salts can be made from acids. Different acids make different types of salt.

- Hydrochloric acid makes chlorides.
- Nitric acid makes nitrates.
- Sulfuric acid makes sulfates.

Several types of substance can supply metal ions to salts, including:

- metals, such as magnesium
- insoluble bases, such as copper oxide
- alkalis, such as sodium hydroxide.

Making a soluble salt from a metal and an acid

Soluble salts are salts that dissolve in water.

You can make some soluble salts by reacting a metal with an acid. For example, to make magnesium sulfate;

- keep adding small pieces of magnesium to sulfuric acid until there is no more bubbling and a little solid magnesium remains
- filter to remove unreacted magnesium
- heat the solution over a water bath, until about half its water has evaporated
- leave the solution to stand for a few days. Magnesium sulfate crystals will form. This is **crystallisation**.

The equation for the reaction is:

magnesium + sulfuric acid → magnesium sulfate + hydrogen

$$Mg(s) + H_2SO_4(aq) \rightarrow MgSO_4(aq) + H_2(g)$$

You can't make all metal salts like this. Some metals, such as copper, are not reactive enough. Some, like sodium, are too reactive – it is not safe to add sodium metal to dilute acids in a school science laboratory.

Exam tip AQA

Remember – hydrochloric acid makes chloride salts, sulfuric acid makes sulfates, and nitric acid makes nitrates.

Making a soluble salt from an insoluble base and an acid

Key words

salt, soluble, crystallisation

You can make some metal salts by reacting an insoluble base with an acid. For example, to make copper sulfate:

1 add copper oxide to sulfuric acid until some copper oxide remains unreacted
2 filter the mixture to remove the unreacted copper oxide
3 heat the solution over a water bath, until about half the water has evaporated
4 leave the solution to stand for a few days. Copper sulfate will crystallise from the solution.

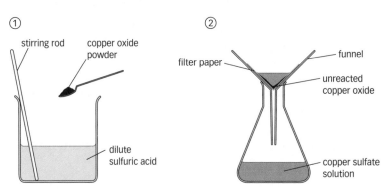

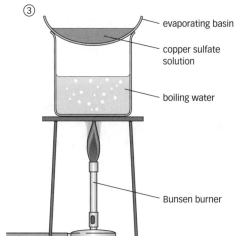

The equation for the reaction is:

copper oxide + sulfuric acid → copper sulfate + water

$$CuO(s) \quad + \quad H_2SO_4(aq) \quad \rightarrow \quad CuSO_4(aq) \quad + H_2O(l)$$

Making a soluble salt from an acid and an alkali

The pictures below show how to make sodium chloride from an acid and an alkali. The equation for the reaction is:

sodium hydroxide + hydrochloric acid → sodium chloride + water

$$NaOH(aq) + \quad HCl(aq) \quad \rightarrow NaCl(aq) + H_2O(l)$$

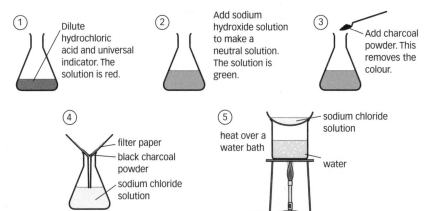

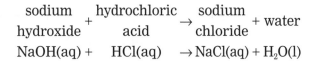

You can use different pairs of acids and alkalis to make different salts.

Questions

1 What is a salt?

2 Describe how to make zinc sulfate from zinc and sulfuric acid.

3 Name the salt made by reacting zinc oxide with hydrochloric acid. Write a word equation for the reaction.

4 Describe how to make crystals of potassium nitrate from potassium hydroxide solution and an acid. Write a word equation for the reaction.

Revision objectives

- ✔ predict the solutions needed to make an insoluble salt
- ✔ give an example of how precipitation reactions are useful

Making insoluble salts

You can make insoluble salts from solutions in **precipitation** reactions. For example, reacting lead nitrate solution with potassium iodide solution produces a **precipitate** of lead iodide. A precipitate is a suspension of small solid particles, spread throughout a liquid or solution. It makes the liquid look cloudy.

$$\text{lead nitrate} + \text{potassium iodide} \rightarrow \text{lead iodide} + \text{potassium nitrate}$$

$$Pb(NO_3)_2(aq) + 2KI(aq) \rightarrow PbI_2(s) + 2KNO_3(aq)$$

The equation below shows only the ions that take part in the reaction. It is an **ionic equation**.

$$Pb^{2+}(aq) + 2I^-(aq) \rightarrow PbI_2(s)$$

The starting solutions are colourless. Lead iodide is bright yellow.

You can separate solid lead iodide from the potassium nitrate solution by filtering the mixture.

Predicting precipitates

To predict how to make insoluble salts, you need to know which salts are soluble in water.

Salts	Are they soluble?
nitrates	all soluble
chlorides	all soluble, expect for lead chloride and silver chloride
sulfates	all soluble, expect for lead sulfate, calcium sulfate, and barium sulfate

Lithium, sodium, and potassium salts are also soluble.

Using precipitation reactions

Precipitation reactions are used to remove unwanted ions from solutions, for example, in treating water for drinking.

Questions

1   What is a precipitation reaction?

2   Suggest **two** soluble salts you could use to make a precipitate of barium sulfate.

1 Draw lines to match each state symbol to its meaning.

state symbol
(aq)
(s)
(l)
(g)

meaning
gas
dissolved in water
solid
liquid

2 In the sentences below, highlight the one bold word or phrase that is correct.

   a Hydrochloric acid makes **chlorides/ hydrochlorides**.

   b Nitric acid makes **nitrics/nitrates**.

   c Sulfuric acid makes **sulfides/sulfates**.

3 Choose words from the box below to fill in the gaps in the sentences that follow. The words in the box may be used once, more than once, or not at all.

> ammonium   acidic   fertilisers   sodium   drinks
> ammonia   alkaline

Ammonia dissolves in water to produce an _____ solution. It is used to produce _____ salts. These salts are important _____.

4 Highlight the statements below that are true. Then write corrected versions of the statements that are false.

   a Crystallisation is the formation of salt crystals from a salt solution.

   b Bases are oxides of non-metals.

   c Sulfur dioxide is a base.

   d Zinc oxide is a base.

   e Hydroxide ions make solutions alkaline.

   f Soluble hydroxides are called alkalis.

   g Potassium hydroxide is an alkali.

   h Magnesium oxide is a base.

   i Hydrogen ions make solutions alkaline.

   j The pH of a neutral solution is 7.

   k A solution with a pH of 6 is alkaline.

5 The statements below describe the steps in making a salt from magnesium and hydrochloric acid. Write the letters of the steps in the best order.

   A Filter the mixture.

   B Place the evaporating basin over a beaker of boiling water.

   C Add magnesium to the acid until some solid magnesium remains.

   D Leave in a warm place for a few days.

   E Pour the filtrate into an evaporating basin.

   F Heat until about half the water has evaporated from the solution.

6 Complete the table below.

Solution type	pH
acidic	_____ than 7
neutral	
alkaline	_____ than 7

7 Name the soluble salt made from each pair of substances below.

   a Magnesium and hydrochloric acid.

   b Copper oxide and sulfuric acid.

   c Potassium hydroxide and nitric acid.

   d Magnesium oxide and sulfuric acid.

8 Complete the neutralisation equation below.

$$H^+ \underline{\quad} + \underline{\quad} (aq) \rightarrow \underline{\quad} (l)$$

9 State whether the solutions in the table are acidic, alkaline or neutral.

Solution contains...	Acidic, alkaline or neutral?
an equal number of OH⁻ and H⁺ ions.	
more H⁺ ions than OH⁻ ions.	
more OH⁻ ions than H⁺ ions.	

10 Write instructions for making magnesium sulfate crystals from magnesium oxide and sulfuric acid.

11 Write instructions for making sodium chloride crystals from an acid and an alkali in a neutralisation reaction. Include the names of the starting materials.

12 Fill in the gaps in the sentences below to describe how to make solid lead iodide from two solutions. Place a solution of lead _____ in a beaker. Add a solution of _____ iodide. A bright yellow _____ forms. Filter the mixture. Solid _____ _____ remains in the filter paper. The filtrate is a solution of _____ _____.

13 Name the insoluble salt made from each pair of solutions below:

   a lead nitrate and potassium iodide

   b barium chloride and sodium sulfate

   c lead nitrate and sodium iodide

   d silver nitrate and sodium hydroxide

   e barium nitrate and lithium sulfate.

14 Suggest pairs of solutions that could be used to make the insoluble salts listed below:

   a lead chloride

   b calcium sulfate

   c lead sulfate

   d barium sulfate.

1 A student wanted to make zinc chloride crystals by reacting a metal with an acid.

a Which acid should the student use?

Tick (✓) **one** box.

Acid	Tick (√)
hydrochloric acid	
nitric acid	
sulfuric acid	

(1 mark)

b The student placed some acid in a conical flask.

He added zinc metal until no more would react.

He filtered the resulting mixture.

Explain why the student filtered the mixture.

....

....

(1 mark)

c The student poured the zinc chloride solution into an evaporating dish.

He heated it over a water bath until half the solution had evaporated.

He placed the dish and its contents in a warm place.

A week later, the student observed crystals in the dish.

Name the process by which zinc chloride crystals formed from the zinc chloride solution.

Tick (✓) **one** box.

Name of process	Tick (✓)
filtration	
neutralisation	
solution	
crystallisation	

(1 mark)

d The hazard symbols below appear on a bottle of zinc chloride crystals.

Suggest **two** ways of managing the risks from these hazards when dealing with the zinc chloride made in the experiment.

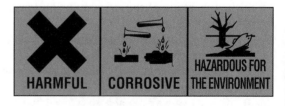

1 ....

2 ....

(2 marks)
*(**Total marks: 5**)*

2 The table gives data about four oxides.

Name of oxide	Does it dissolve in water?
copper oxide	no
magnesium oxide	no
sodium hydroxide	yes
carbon dioxide	yes

a Use data from the periodic table and the table above to name:

i **three** bases

....
(1 mark)

ii **one** alkali

....
(1 mark)

b Give the name and formula of the ion that makes solutions alkaline.

Name ....

Formula ....
(2 marks)
(Total marks: 4)

3 A student plans to make solid lead iodide in a precipitation reaction.

a Name **two** solutions that the student could use.

1 ....

2 ....
(2 marks)

b Describe how the student could make solid lead iodide from the two solutions you named in part a.

....

....
(2 marks)

c The equation below shows the ions that take part in the precipitation reaction to make lead iodide.

Complete the equation by adding state symbols for each substance shown in the equation.

$$Pb^{2+} .... + 2\ I^- .... \rightarrow PbI_2 ....$$
(3 marks)
(Total marks: 7)

Revision objectives

- describe what happens at the electrodes in electrolysis
- use half equations to represent reactions at electrodes

Student book references

2.29 Electrolysis

Specification key

✓ C2.7.1 a – c, e – g

What is electrolysis?

When an ionic substance is molten, or dissolved in water, its ions can move about.

Passing an electric current through the liquid or solution breaks down the ionic compound into simpler substances. This is **electrolysis**. The substance that is broken down is the **electrolyte**.

During electrolysis, the ions move to the **electrodes**:
- Positive ions move to the negative electrode.
- Negative ions move to the positive electrode.

This diagram shows what happens when an electric current passes through molten lead bromide.

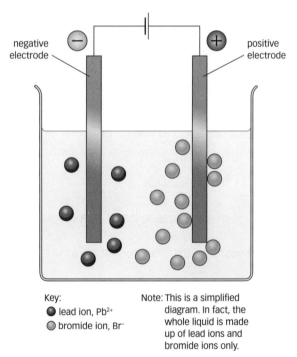

Key:
● lead ion, Pb^{2+}
○ bromide ion, Br^-

Note: This is a simplified diagram. In fact, the whole liquid is made up of lead ions and bromide ions only.

▲ The electrolysis of molten lead bromide.

In the lead bromide electrolysis cell two things happen:
- Positively charged lead ions move towards the negative electrode. Lead metal forms here.
- Negatively charged bromide ions move towards the positive electrode. Liquid bromine forms here.

What happens at the electrodes?

- At the negative electrode, positively charged ions gain electrons. This is **reduction**. In the example, lead ions gain electrons. The positive ions are reduced.
- At the positive electrode, negatively charged ions lose electrons. This is **oxidation**. The negative ions are oxidised.

H Half equations

Chemists use **half equations** to show what happens at the electrodes. In half equations, an electron is represented as e⁻. You can balance half equations by adding or subtracting electrons until the total charge on each side is equal.

For example, in the electrolysis of lead bromide:
- at the negative electrode, $Pb^{2+} + 2e^- \rightarrow Pb$
- at the positive electrode, $Br^- \rightarrow Br + e^-$

This is also written as $Br^- - e^- \rightarrow Br$

The bromine atoms then join together in pairs to make bromine molecules:
$$Br + Br \rightarrow Br_2$$

Predicting products

If you electrolyse molten lead bromide, there is only one possible product at each electrode. But for compounds that are dissolved in water, there are other possible products, since water also takes part in electrolysis reactions.

The table below shows the products when electricity passes through some solutions.

Solution	Product at negative electrode	Product at positive electrode
copper chloride	copper	chlorine
magnesium iodide	hydrogen	iodine
silver nitrate	silver	oxygen
sodium sulfate	hydrogen	oxygen
potassium carbonate	hydrogen	oxygen

At the negative electrode:
- the metal is produced if it is low in the reactivity series
- hydrogen is produced if the metal is above copper in the reactivity series. The hydrogen is formed by the electrolysis of water.

At the positive electrode:
- halogens are produced if there are halide ions in the solution
- oxygen is produced if there are nitrate, sulfate or carbonate ions in the solution. The oxygen comes from the water.

Exam tip

Remember OIL RIG – Oxidation Is Loss (of electrons) and Reduction Is Gain (of electrons).

Questions

1 Name the products formed at the positive and negative electrodes in the electrolysis of molten copper chloride.

2 Predict the products formed at the electrodes during the electrolysis of a solution of sodium carbonate in water.

3 H Write half equations for the reactions that occur at the electrodes during the electrolysis of molten copper chloride.

Revision objectives

- ✔ describe how electroplating works
- ✔ describe and explain how aluminium is extracted from aluminium oxide by electrolysis
- ✔ name the products of the electrolysis of sodium chloride solution

Student book references

2.30 Electroplating
2.31 Using electrolysis – 1
2.32 Using electrolysis – 2

Specification key

✔ C2.7.1 d, h – i

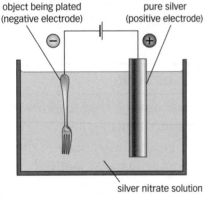

▲ This diagram shows how to electroplate an object with silver.

object being plated (negative electrode)
pure silver (positive electrode)
silver nitrate solution

Electroplating

Electrolysis is used to coat objects with a thin layer of a metal such as copper, silver or tin. This is **electroplating**. Electroplating has the following two main purposes:

- to protect a metal object from corrosion by coating it with an unreactive metal that does not easily corrode
- to make an object look attractive.

Food cans are made of tinplate – steel that has been electroplated with tin. Cutlery may be electroplated with silver.

Extracting aluminium

Aluminium is extracted from **bauxite ore**. Bauxite is mainly aluminium oxide. Pure aluminium is obtained from aluminium oxide in the following way.

Dissolve aluminium oxide in molten **cryolite**. The solution formed has a lower melting point than pure aluminium oxide. So less energy is needed to keep the mixture liquid than would be needed for aluminium oxide alone.

Pour the aluminium oxide and cryolite mixture into a huge electrolysis cell.

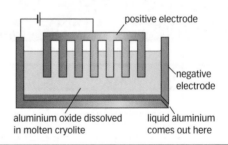

positive electrode
negative electrode
aluminium oxide dissolved in molten cryolite
liquid aluminium comes out here

Pass an electric current through the liquid mixture.

Aluminium ions move to the negative electrode. They gain electrons to make liquid aluminium metal.

H $Al^{3+} + 3e^- \rightarrow Al$

Oxide ions move to the positive electrode. They lose electrons to form oxygen atoms, which join together in pairs to make oxygen gas. The oxygen gas reacts with the carbon of the electrode and forms carbon dioxide gas.

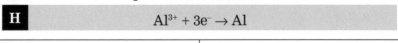

H $O^{2-} \rightarrow O + 2e^-$ then $O + O \rightarrow O_2$ then $C + O_2 \rightarrow CO_2$

The electrolysis of sodium chloride solution

The electrolysis of concentrated sodium chloride solution, or **brine**, produces several products:

• Hydrogen gas forms at the negative electrode.
• Chlorine gas forms at the positive electrode.

A solution of sodium hydroxide also forms during the process.

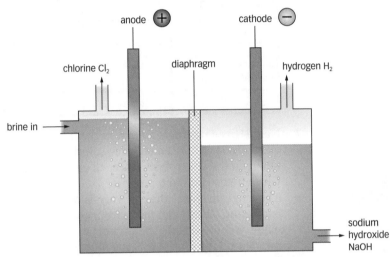

▲ The apparatus used in the industrial electrolysis of brine.

The products have many uses:

• Sodium hydroxide is used to make soap.
• Chlorine is used to make bleach and plastics, and to sterilise water.
• Hydrogen is used to make margarine and ammonia.

Exam tip

Remember, in electroplating the object needs to be placed at the negative electrode.

Questions

1 Give **two** reasons for electroplating objects.

2 Name the **two** products formed in the industrial electrolysis of aluminium oxide.

3 Name the substances formed at each electrode in the electrolysis of sodium chloride solution.

Working to Grade E

1 Tick **two** possible reasons for electroplating an object.
  a To protect the object from corrosion. ☐
  b To make the object look more attractive. ☐
  c To make the object harder to detect with a metal detector. ☐

2 Choose words from the box below to fill in the gaps in the sentences that follow. The words in the box may be used once, more than once, or not at all.

> solid evaporated melted atoms ions
> solution dissolved

When an ionic substance is _____ or _____ in water, the _____ are free to move about within the liquid or _____.

3 Draw lines to match each word to its definition.

Word	Definition
electrolyte	A liquid or solution that is broken down when electricity passes through it.
electrolysis	Covering an object with a layer of a metal in an electrolysis cell.
electroplating	Pieces of metal or graphite through which electricity enters or leaves an electrolysis cell.
electrodes	The process by which electricity breaks down a liquid or solution.

4 Highlight the correct word in each pair of bold words.
Molten lead bromide is made up of negatively charged **bromide/bromine** ions and **negatively/positively** charged lead ions. A student passes electricity through molten lead bromide. The negative ions move to the **positive/negative** electrode. The lead ions move to the **positive/negative** electrode.

Working to Grade C

5 Highlight the statements below that are true. Then write corrected versions of the sentences that are false.
  a At the negative electrode, positively charged ions lose electrons.
  b At the positive electrode, negatively charged ions lose electrons.
  c If an ion gains electrons, the ion is oxidised.
  d Reduction occurs when an ion gains electrons.

6 Predict what is formed at the negative and positive electrodes when electricity passes through the solutions in the table.

Solution	Positive electrode	Negative electrode
copper chloride		
potassium bromide		
silver nitrate		
magnesium nitrate		
copper carbonate		
sodium sulfate		

7 Name **three** products formed in the industrial electrolysis of sodium chloride solution. State where in the electrolysis cell each one is produced. Give **one** use for **each** product.

8 The diagram shows the electrolysis cell for extracting aluminium metal from aluminium oxide. Write each label below next to the correct number on the diagram. Some numbers on the diagram have more than one label.

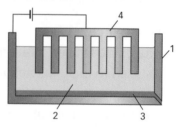

Labels:
A Positive electrode
B Positive ions gain electrons at this electrode
C Negative electrode
D Electrolyte of liquid aluminium and cryolite
E Liquid aluminium forms at this electrode
F Oxide ions are oxidised at this electrode
G Carbon dioxide gas forms here
H This electrode is made of carbon
I Reduction happens at this electrode
J Negative ions are attracted to this electrode

Working to Grade A*

9 Write half equations for the reactions that happen at the positive and negative electrodes during the electrolysis of:
  a Molten lead bromide, $PbBr_2$
  b Copper chloride solution, $CuCl_2$
  c Molten aluminium oxide, Al_2O_3

1 Some silver jewellery is electroplated with a layer of rhodium.

a The table gives some data for silver and rhodium.

	silver	rhodium
relative hardness	25	100
melting point (°C)	961	1970
colour	silvery-white	silvery-white
corrosion	reacts with hydrogen sulfide in the air to make black silver sulfide	does not react with gases in the air

Use the data in the table to help you suggest two reasons for electroplating silver jewellery.

1 ....

2 ....

(2 marks)

b The diagram below shows an electrolysis cell used to electroplate a silver ring.

The electrolyte contains positively charged rhodium ions, Rh^{3+}.

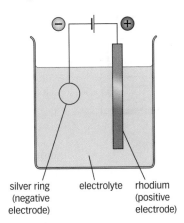

silver ring   electrolyte   rhodium
(negative   (positive
electrode)   electrode)

i The silver ring is used as the negative electrode in the electrolysis cell. Explain why.

....

(1 mark)

ii At the negative electrode, rhodium ions gain electrons to become rhodium atoms.

What type of reaction is this?

Put a ring around the one correct answer.

       corrosion   **oxidation**   **reduction**   **electrolysis**   *(1 mark)*

H   **iii** Write a half equation to represent the reaction that happens at the negative electrode.

....

(1 mark)

(Total marks: 5)

Designing an investigation, presenting data, and using data to draw conclusions

Skill – Understanding the experiment

1   A student investigates the hypothesis *There is a relationship between the rate of a reaction and temperature*. She sets up the apparatus below.

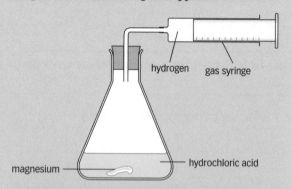

magnesium — hydrogen — gas syringe — hydrochloric acid

The student's results are in the table.

Temp (°C)	Time to collect 50 cm³ of hydrogen gas (s)			
	Test 1	Test 2	Test 3	Mean
20	401	401	404	402
30	196	197	195	196
40	376	100	100	192
50	47	47	50	
60	25	25	25	25

a   Identify the dependent variable, the independent variable and **three** control variables for the investigation.

To answer this question, you need to know that the independent variable is the one that the student changes. The dependent variable is the one that the student measures for each change in the independent variable.

The control variables are the variables that the student keeps the same – these are not given in the question. You will need to use your knowledge and experience of doing investigations, and the diagram, to help you answer this part of the question.

Skill – Using data to draw conclusions

b   Calculate the missing mean.

The mean is the sum of the measurements divided by the number of measurements taken. For example, the mean of the time values at 20 °C is:
(401 s + 401 s + 404 s) ÷ 3 = 402 s

c   Plot the results on a graph.

Both the independent and the dependent variables are continuous – they can take any numerical value. This means that you can draw a line graph to display the data. Don't forget to draw a line of best fit rather than joining up all the points.

d   Do the results support the hypothesis? Explain why.

To answer this question you will need to look back at the hypothesis at the start of the question. If the line graph you plotted in part c shows the relationship given in the hypothesis, then the results support the hypothesis.

Societal aspects of scientific evidence

Skill – Analysing the facts and making deductions

2   Making ibuprofen

In the past, an aluminium chloride catalyst was used in the manufacture of the painkiller ibuprofen. This catalyst could not be separated from the reaction mixture, and so could not be reused.

Today, two other catalysts are used in the production of ibuprofen – hydrogen fluoride and a nickel/aluminium alloy. These catalysts can be used again and again.

a   Suggest an economic benefit of reusing a catalyst, compared to using a fresh sample of catalyst for each batch.

Economic factors are those to do with money, so you will need to explain why reusing a catalyst might help to increase the profits of a company making ibuprofen.

Skill – Understanding the impact of a decision

b   Suggest **two** environmental benefits of reusing a catalyst.

Environmental factors might be linked to energy requirements (for example, those to extract raw materials, or to manufacture a product). Environmental factors are also linked to waste disposal. For example, how are unwanted by-products of an industrial process disposed of? What happens to a product once it has done its job, and is no longer useful?

AQA Upgrade

Answering an extended writing question

1 *In this question you will be assessed on using good English, organising information clearly, and using specialist terms where appropriate.*
Hydrochloric acid reacts with calcium carbonate to make calcium chloride, water, and carbon dioxide gas. Describe and explain the factors that affect the rate of this reaction. *(6 marks)*

G–E

if U like make it more conc then it is faster becoz like the particals hit each other lots more often and there is a gass so the preshure going up makes it fasster and if you make it collder it shoud speed up I reckon

Examiner: This answer is typical of a grade-G candidate. It is worth just one mark, gained for describing and explaining that increasing the acid concentration increases the reaction rate. The candidate is incorrect in stating that increasing the pressure increases the rate.

The candidate has not used specialist terms. There is no punctuation, and there are several spelling mistakes.

D–C

If temp go up so does the rate becuase the particals move faster.
also if you have powder with big surface area it is faster, and can bubble up right to the top of the flask and even overflow all over everywhere and make a mess very quickly and the teacher might get annoyed because it is dangerous.
With dilute acid it is slower than with consentrated acid.

Examiner: This answer is worth three marks out of six. It is typical of a grade-C or -D candidate. The candidate has described three factors that affect reaction rate, and partially explained one of these in terms of particles and collision theory.

The answer is well organised, with a few mistakes of grammar and punctuation. There are spelling mistakes. Part of the answer is not relevant.

B–A*

The rate can be affected by acid concentration, temperature and surface area of the calcium carbonate.
Increasing the acid concentration increases the rate. This is because there are more acid particles in a certain volume of solution, so they can collide more frequently with the calcium carbonate.
Increasing surface area of the calcium carbonate increases the rate because with a powder there are more particles exposed and ready to react. So the collisions are more frequent and the rate goes up.
Increasing temperature means the particles move faster. So they collide more often. They also collide with more energy. This increases the rate.

Examiner: This is a high-quality answer, typical of an A* candidate. It is worth six marks out of six.

The question has been answered accurately and the explanations are detailed.

The answer is well organised, and includes a short introduction. The spelling, punctuation, and grammar are faultless. The candidate has used several specialist terms.

2 *In this question you will be assessed on using good English, organising information clearly, and using specialist terms where appropriate.*

Describe how to make copper sulfate crystals from an insoluble metal oxide and a dilute acid.

(6 marks)

G–E

Mix up hydraulic acid and Copper Oxsyde. Heat it all up? Stir and filter! You've got xtals.

Examiner: This answer is worth one mark out of six, for correctly identifying the starting materials. It is typical of a grade-F or -G candidate.

There are several spelling and punctuation mistakes, and the stages described are not in the correct order.

D–C

Add copper oxide powder to hydrochloric acid. Stir. You adds a bit more. Then you filters it all. Keep the blue liquid. Heat it up and you made crystals round the edge of the dish.

Examiner: This answer is worth three marks out of six, and is typical of a grade-C or -D candidate. The candidate has given one incorrect starting material, and has described the stages in insufficient detail. The candidate has not described the final stage of the preparation of the crystals.

The answer is well organised. The spelling and punctuation are accurate, but there are some grammatical errors.

B–A*

Place some dilute sulfuric acid in a beaker. Add copper oxide powder with a spatula, and stir, until no more will dissolve. Then filter to remove the unreacted black solid from the blue solution. Pour the blue solution – the filtrate – into an evaporating basin. Heat over a water bath until half the water has evaporated. Now the solution is more concentrated.
Then leave the solution in a warm place for a week. Crystals will form.

Examiner: This answer gains six marks out of six, and is typical of an A or A* candidate. It describes in detail, using scientific words and correct apparatus names, how to make the salt from correctly identified starting materials.

The answer is logically organised, and includes scientific words. The spelling, punctuation and grammar are accurate.

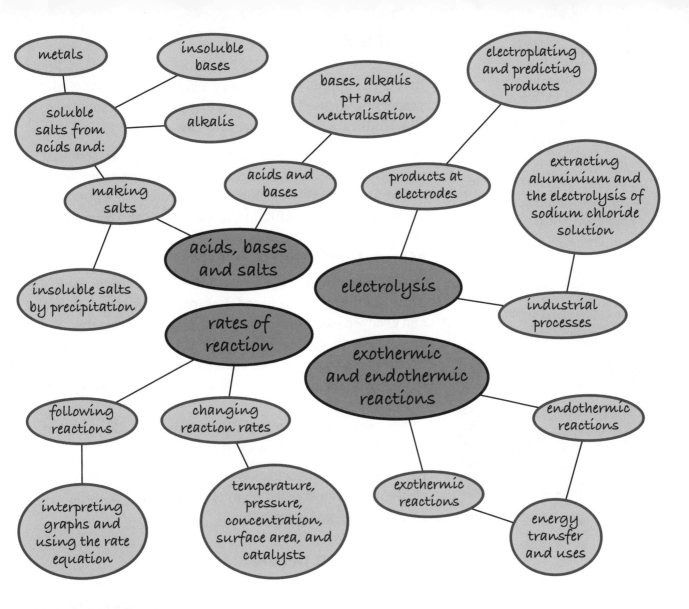

Revision checklist

- You can find the rate of a chemical reaction by measuring the amount of a product used or the amount of a product formed over time.

- Rate of reaction = $\dfrac{\text{amount of reactant used}}{\text{time}}$

  or $\dfrac{\text{amount of product made}}{\text{time}}$

- Chemical reactions only happen when reacting particles collide with each other and with enough energy.

- Increase the rate of a reaction by increasing temperature, or pressure (of reacting gases), or concentration (of reacting solutions), or surface area (of reacting solids) or by adding a catalyst.

- Exothermic reactions transfer energy to the surroundings.

- Endothermic reactions take in energy from the surroundings.

- Soluble salts are made from acids by reacting them with metals, or insoluble bases, or alkalis. The salt solutions can be crystallised to make solid salts.

- Different acids produce different salts – hydrochloric acid makes chlorides, nitric acid makes nitrates, sulfuric acid makes sulfates.

- Insoluble salts are made in precipitation reactions.

- Hydrogen ions (H^+) make solutions acidic and hydroxide ions (OH^-) make solutions alkaline. These ions react together in neutralisation reactions.

- Ionic substances conduct electricity when liquid or dissolved in water since their ions are free to move.

- In electrolysis, positive ions are attracted to the negative electrode. Here they gain electrons. This is reduction.

- Negative ions are attracted to the positive electrode. Here they lose electrons. This is oxidation.

- Aluminium is extracted from aluminium oxide by the electrolysis of a molten mixture of aluminium oxide and cryolite. Aluminium forms at the negative electrode. Carbon dioxide forms at the positive electrode, which is made of carbon.

- The electrolysis of sodium chloride solution produces hydrogen at the negative electrode and chlorine at the positive electrode. Sodium hydroxide solution is also formed.

Revision objectives

- ✔ evaluate the work of Mendeleev and Newlands
- ✔ describe the characteristics of the modern periodic table
- ✔ link electronic structure to position in the periodic table
- ✔ describe the properties of the transition metals

Student book references

3.1 Organising the elements

3.2 The periodic table

3.5 The transition elements

Specification key

✔ C3.1.1 ✔ C3.1.2

✔ C3.1.3 c – d

Newlands octaves

In the 1860s, John Newlands listed the 56 elements then known in order of atomic weight. Every eighth element had similar properties. He used this pattern to group the elements. Newlands called his discovery the **law of octaves**.

Other chemists criticised the law of octaves. They said that some groups included an element whose properties were very different to the other elements in the group. For example, nickel was grouped with fluorine, chlorine, and bromine.

Mendeleev's periodic table

In 1869 Dmitri Mendeleev created the first **periodic table**. He arranged the elements in order of increasing atomic weight, and grouped elements with similar properties together. But some elements seemed to be in the wrong groups. He overcame this problem by:

- swapping the positions of some elements, for example, iodine and tellurium
- leaving gaps for elements that he predicted did exist, but that had not been discovered.

Later, other scientists discovered some of the missing elements. This increased scientists' confidence that the periodic table was a useful tool.

The modern periodic table

Early in the twentieth century, scientists discovered the particles that make up atoms – protons, neutrons, and electrons.

1	2											3	4	5	6	7	0
						1 **H** Hydrogen 1											4 **He** Helium 2
7 **Li** Lithium 3	9 **Be** Beryllium 4											11 **B** Boron 5	12 **C** Carbon 6	14 **N** Nitrogen 7	16 **O** Oxygen 8	19 **F** Fluorine 9	20 **Ne** Neon 10
23 **Na** Sodium 11	24 **Mg** Magnesium 12											27 **Al** Aluminium 13	28 **Si** Silicon 14	31 **P** Phosphorus 15	32 **S** Sulfur 16	35.5 **Cl** Chlorine 17	40 **Ar** Argon 18
39 **K** Potassium 19	40 **Ca** Calcium 20	45 **Sc** Scandium 21	48 **Ti** Titanium 22	51 **V** Vanadium 23	52 **Cr** Chromium 24	55 **Mn** Manganese 25	56 **Fe** Iron 26	59 **Co** Cobalt 27	59 **Ni** Nickel 28	63.5 **Cu** Copper 29	65 **Zn** Zinc 30	70 **Ga** Gallium 31	73 **Ge** Germanium 32	75 **As** Arsenic 33	79 **Se** Selenium 34	80 **Br** Bromine 35	84 **Kr** Krypton 36
85 **Rb** Rubidium 37	88 **Sr** Strontium 38	89 **Y** Yttrium 39	91 **Zr** Zirconium 40	93 **Nb** Niobium 41	96 **Mo** Molybdenum 42	[98] **Tc** Technetium 43	101 **Ru** Ruthenium 44	103 **Rh** Rhodium 45	106 **Pd** Palladium 46	108 **Ag** Silver 47	112 **Cd** Cadmium 48	115 **In** Indium 49	119 **Sn** Tin 50	122 **Sb** Antimony 51	128 **Te** Tellurium 52	127 **I** Iodine 53	131 **Xe** Xenon 54
133 **Cs** Caesium 55	137 **Ba** Barium 56	139 **La*** Lanthanum 57	178 **Hf** Hafnium 72	181 **Ta** Tantalum 73	184 **W** Tungsten 74	186 **Re** Rhenium 75	190 **Os** Osmium 76	192 **Ir** Iridium 77	195 **Pt** Platinum 78	197 **Au** Gold 79	201 **Hg** Mercury 80	204 **Tl** Thallium 81	207 **Pb** Lead 82	209 **Bi** Bismuth 83	[209] **Po** Polonium 84	[210] **At** Astatine 85	[222] **Rn** Radon 86
[223] **Fr** Francium 87	[226] **Ra** Radium 88	[227] **Ac*** Actinium 89	[261] **Rf** Rutherfordium 104	[262] **Db** Dubnium 105	[266] **Sg** Seaborgium 106	[264] **Bh** Bohrium 107	[277] **Hs** Hassium 108	[268] **Mt** Meitnerium 109	[271] **Ds** Darmstadtium 110	[272] **Rg** Roentgenium 111							

Key
relative atomic mass
atomic symbol
name
atomic (proton) number

Elements with atomic numbers 112–116 have been reported but not fully authenticated

* The Lanthanides (atomic numbers 58–71) and the Actinides (atomic numbers 90–103) have been omitted.
Cu and **Cl** have not been rounded to the nearest whole number.

▲ The modern periodic table. It is called a periodic table because elements with similar properties occur at regular intervals. Elements with similar properties are in columns, called groups. The horizontal rows are called periods.

The scientists arranged the elements in order of increasing **atomic number** (proton number) to make a new periodic table. All the elements were now in appropriate groups. The problems of Mendeleev's periodic table, based on atomic weights, had been solved.

Electronic structure and the periodic table

An element's position in the periodic table is linked to its electronic structure. Elements in the same group have the same number of electrons in their highest occupied energy level (outer shell).

For the main groups, the number of electrons in the highest occupied energy level is equal to the group number.

The transition elements

The **transition elements** form the central block of the periodic table.

the transition elements

												H							He
Li	Be												B	C	N	O	F	Ne
Na	Mg												Al	Si	P	S	Cl	Ar
K	Ca	Sc	Ti	V	Cr	Mn	Fe	Co	Ni	Cu	Zn		Ga	Ge	As	Se	Br	Kr
Rb	Sr	Y	Zr	Nb	Mo	Tc	Ru	Rh	Pd	Ag	Cd		In	Sn	Sb	Te	I	Xe
Cs	Ba	La	Hf	Ta	W	Re	Os	Ir	Pt	Au	Hg		Tl	Pb	Bi	Po	At	Rn
Fr	Ra	Ac	Rf	Db	Sg	Bh	Hs	Mt	Ds	Rg								

The transition elements are metals. Most transition elements:

- are strong and hard
- have high densities
- have high melting points (except for mercury, which is liquid at room temperature).

The transition elements react slowly – if at all – with water and oxygen:

- Platinum and gold do not react with water and oxygen.
- Iron reacts slowly with water and oxygen at room temperature to make rust (hydrated iron(III) oxide).

Many transition elements form ions with different charges. For example, copper has two oxides:

- CuO is black. It includes a Cu^{2+} ion.
- Cu_2O is red. It includes a Cu^+ ion.

Many transition elements form coloured compounds. For example, vanadium compounds can be yellow, blue, green, or lilac.

Transition elements and their compounds are useful catalysts. For example, vanadium pentoxide (V_2O_5) speeds up a vital step in the manufacture of sulfuric acid.

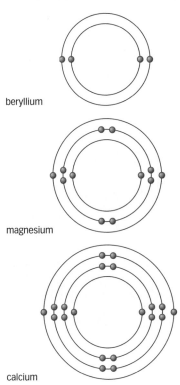

beryllium

magnesium

calcium

▲ The Group 2 elements all have two electrons in their highest occupied energy level.

Exam tip

Remember, groups are vertical columns, and periods are horizontal rows.

Questions

1 Describe how Newlands organised the elements.

2 Identify **two** ways in which Mendeleev's organisation of the elements was better than Newlands'.

3 List **seven** properties that are typical of the transition elements.

Revision objectives

- ✔ compare the properties of Group 1 elements and transition elements
- ✔ describe the reactions of the Group 1 elements with non-metals
- ✔ explain the trend in reactivity of the Group 1 elements

Student book references

3.3 Alkali metals – 1

3.4 Alkali metals – 2

Specification key

✔ C3.1.3 a – b and h (part)

Physical properties

The Group 1 elements are called the **alkali metals.** The group includes lithium, sodium, and potassium.

Group 1
the alkali metals

| Li | Be | | | | | | | | | | | B | C | N | O | F | Ne |

▲ Group 1 is on the left of the periodic table.

The alkali metals are soft – you can cut them with a knife. They also have low densities. The densities of lithium, sodium, and potassium are so low that they can float on water.

The alkali metals have low melting and boiling points compared to all transition elements except mercury. The further down Group 1 an element is, the lower its melting and boiling points.

Name of element	Melting point (°C)	Boiling point (°C)
lithium	180	1330
sodium	98	890
potassium	64	774
rubidium	39	688

Reactions with non-metals

The alkali metals react with non-metals such as chlorine. For example, sodium burns vigorously in chlorine. The product is sodium chloride (common salt).

$$\text{sodium} + \text{chlorine} \rightarrow \text{sodium chloride}$$
$$2Na\,(s) + Cl_2\,(g) \rightarrow 2NaCl\,(s)$$

The alkali metals also react with oxygen, another non-metal.

- At room temperature, their surfaces tarnish when exposed to air. This is why they are stored in oil.
- On heating, they react vigorously with oxygen from the air. For example:

$$\text{sodium} + \text{oxygen} \rightarrow \text{sodium oxide}$$
$$4Na\,(s) + O_2\,(g) \rightarrow 2Na_2O\,(s)$$

Alkali metal compounds

A compound made up of an alkali metal and a non-metal is **ionic**. The metal ion has a charge of +1. For example:

- Potassium chloride (KCl) is made up of potassium ions (K^+) and chloride ions (Cl^-). In a crystal of the compound, there is one potassium ion for every chloride ion.
- Sodium oxide (Na_2O) is made up of sodium ions (Na^+) and oxide ions (O^{2-}). In a crystal of the compound, there are two sodium ions for every one oxide ion.

The compounds are white solids at room temperature. They dissolve in water to form colourless solutions.

Reactions with water

The alkali metals react vigorously with water. The products are:

- hydrogen gas
- a hydroxide.

For example:

sodium + water → sodium hydroxide + hydrogen

$$2Na\,(s) + 2H_2O\,(l) \rightarrow 2NaOH\,(aq) + H_2\,(g)$$

Alkali metal hydroxides dissolve in water to give alkaline solutions.

Group trend

The Group 1 elements are more reactive than the transition elements. So their reactions with oxygen and water are more vigorous than those of the transition elements.

The further down Group 1 an element is, the more vigorous its reactions.

> **H** This trend can be explained by the energy level of the outer electrons. Potassium is lower down Group 1 than lithium. The outermost electron of potassium is in a higher energy level than that of lithium. This means that, in reactions, potassium gives away its outermost electron more easily than lithium. Potassium is more reactive than lithium.

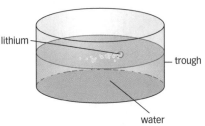

▲ All the alkali metals react vigorously with water. The reactions get more vigorous going down the group. Lithium, at the top, zooms around on the water surface. Caesium, at the bottom, reacts explosively with water.

Questions

1. Draw a table to show **three** differences in the properties of the Group 1 elements and the transition elements.

2. Describe the differences and similarities between the reaction of lithium with water and the reaction of potassium with water.

3. **H** The further down Group 1 an element is, the more vigorous its reactions. Explain why.

Revision objectives

- describe the reactions of the halogens with metals
- describe the displacement reactions of the halogens
- explain the trend in reactivity of the halogens

Student book references

3.6 The halogens

Specification key

✔ C3.1.3 e –g and h (part)

▲ Chlorine molecules are made up of two chlorine atoms joined together by a strong covalent bond.

Physical properties

The Group 7 elements are called the **halogens**. The group includes fluorine, chlorine, bromine, and iodine.

																	Group 7 the halogens	
						H												He
Li	Be											B	C	N	O	F	Ne
Na	Mg											Al	Si	P	S	Cl	Ar
K	Ca	Sc	Ti	V	Cr	Mn	Fe	Co	Ni	Cu	Zn	Ga	Ge	As	Se	Br	Kr
Rb	Sr	Y	Zr	Nb	Mo	Tc	Ru	Rh	Pd	Ag	Cd	In	Sn	Sb	Te	I	Xe
Cs	Ba	La	Hf	Ta	W	Re	Os	Ir	Pt	Au	Hg	Tl	Pb	Bi	Po	At	Rn
Fr	Ra	Ac	Rf	Db	Sg	Bh	Hs	Mt	Ds	Rg							

▲ Group 7 is towards the right of the periodic table.

The Group 7 elements exist as diatomic molecules, for example, Cl_2.

The further down the group an element is, the higher its melting point and boiling points.

Name of element	Melting point (°C)	Boiling point (°C)	State at room temperature	Colour
chlorine	−101	−34.7	gas	green
bromine	−7.20	58.8	liquid	orange/brown
iodine	114	184	solid	grey/black with violet vapour

Reactions with metals

The halogens react with metals to form ionic compounds. For example, chlorine reacts with iron to form iron chloride.

$$iron + chlorine \rightarrow iron\ chloride$$
$$2Fe\ (s) + 3Cl_2\ (g) \rightarrow 2FeCl_3\ (s)$$

Key words

halogen, halide ion

Exam tip

Remember, a more reactive halogen can displace a less reactive halogen from a solution of its salt.

Questions

1 Describe the trends in the melting points and boiling points of the Group 7 elements.

2 Predict which of the reactions below is more vigorous. Give a reason for your decision.
iron + chlorine → iron chloride
iron + iodine → iron iodide

3 Write equations to summarise the displacement reactions of:

a chlorine with sodium iodide solution

b bromine with potassium iodide solution.

Iron chloride is made up of two types of ion:
- iron ions, Fe^{3+}
- chloride ions, Cl^-.

The chloride ion is an example of a **halide ion**. Bromine forms bromide ions (Br^-) and iodine forms iodide ions (I^-). All halide ions have a charge of −1.

In Group 7, the further down the group an element is, the less reactive the element.

H This trend can be explained by the energy levels of the outer electrons in halogen atoms. When iron reacts with chlorine, iron atoms give each chlorine atom one extra electron. The electron completes the outer energy level of the chlorine atom, forming a chloride ion. The higher the energy level of the outer electrons, the less easily electrons are gained.

Displacement reactions

In displacement reactions, a more reactive halogen displaces a less reactive halogen from an aqueous solution of its salt. For example, chlorine is more reactive than bromine, so:

chlorine + potassium bromide → potassium chloride + bromine

$$Cl_2 \, (aq) + 2KBr \, (aq) \rightarrow 2KCl \, (aq) + Br_2 \, (aq)$$

1 Highlight the correct word or phrase in each pair of **bold** words.
Copper is a **Group 1/transition** element. It forms ions with charges that are **different/the same.** Its compounds are **coloured/white.** Copper and its compounds make useful **alkalis/catalysts.**

2 Give the number of electrons in the highest occupied energy level (outer shell) of the elements in:
   a   Group 1   b   Group 3   c   Group 7

3 Write T next to the statement that is true. Write corrected versions of the statements that are false.
   a   In the periodic table, a vertical column is called a period.
   b   In early periodic tables, the elements were arranged in order of atomic weight.
   c   Newlands left gaps in his periodic table for elements he predicted did exist, but had not yet been discovered.

4 Use your knowledge of the properties of the Group 1 elements, and the transition elements, to answer the questions below. You will also need the periodic table. Draw a ring around:
   a   the hardest metal in this list: sodium, manganese, lithium
   b   the metal with the lowest melting point in this list: copper, potassium, gold, chromium
   c   the metal with the highest density in this list: sodium, iron, lithium, potassium
   d   the metal that reacts least vigorously with water in this list: manganese, potassium, caesium, rubidium.

5 Complete the table below.

Name of compound	Formula of metal ion in compound	Appearance and state of compound at room temperature
potassium chloride		
sodium bromide		
lithium chloride		

6 Complete the word equations.
   a   sodium + chlorine → _____
   b   lithium + oxygen → _____
   c   sodium + water → _____ + _____
   d   potassium + water → _____ + _____

7 Decide which of the following pairs of solutions will react in displacement reactions. Then write word equations for the pairs of solutions that react.
   a   chlorine and potassium bromide
   b   iodine and potassium bromide
   c   bromine and potassium iodide
   d   bromine and sodium chloride
   e   chlorine and potassium iodide

8 For the sentences below, write **1** next to each sentence that is true for Group 1. Write **7** next to each sentence that is true for Group 7. To help you, use the data in the table, your own knowledge, and the periodic table.

Element	Boiling point (°C)	Density (g/cm³)
lithium	1330	0.53
sodium	890	0.97
potassium	774	0.86
chlorine	−35	1.56
bromine	59	3.1
iodine	184	4.9

   a   Going down this group, boiling point decreases. ☐
   b   Going down this group, the elements get less reactive ☐
   c   The elements in this group form ionic compounds with non-metals. ☐
   d   The elements in this group form ionic compounds in which the negative ion carries a charge of −1. ☐
   e   The top three elements of this group are less dense than water. ☐
   f   Going down this group, the elements get more reactive. ☐

9 Write balanced symbol equations for the reactions shown by the word equations below.
   a   sodium + bromine → sodium bromide
   b   lithium + water → lithium hydroxide + hydrogen
   c   chlorine + potassium iodide → potassium chloride + iodine

10 Alice has written a paragraph to explain the trend in the reactivities of the Group 7 elements. There is **one** mistake in **each** sentence. Write out the sentences, correcting the mistakes.

When a metal reacts with a halogen, the metal atoms give each halogen atom two extra electrons. This electron completes the inner energy level of the halogen atom. The closer the outer energy level is to the nucleus, the smaller the attraction between the newly added electrons and the nucleus. So the lower down Group 7 an element is, the more easily its atoms gain electrons, and the less reactive the element is.

Examination questions
The periodic table

1 This question is about the periodic table.

a Read the information in the box below. Then answer the questions that follow.

> In 1860 an Italian scientist, Cannizzaro, published a list of the atomic weights of all the elements then known.
>
> In 1869, a Russian scientist, Mendeleev arranged the elements in order of atomic weight. He grouped together elements with similar properties.
>
> He left gaps for elements that he predicted did exist, but that had not been discovered.
>
> In 1875 a French scientist discovered an element to fill one of Mendeleev's gaps. He called it gallium. Four years later, a Swedish scientist discovered another of the missing elements. He called it scandium.

   i Explain how Mendeleev's work relied on the findings of Cannizzaro.

....

....

(1 mark)

   ii Suggest why the discoveries of gallium and scandium in the 1870s made scientists more confident that the periodic table of the time was a useful tool.

....

....

(1 mark)

b **i** Write **two** words to complete the sentence below.

   In the modern periodic table, the elements are arranged in order of their ....

(1 mark)

   ii Explain how, in the modern periodic table, the position of an element in the periodic table is linked to its electronic structure.

   In your answer, include examples and give the electronic structure of at least one element.

....

....

....

(3 marks)
(Total marks: 6)

2 Describe some differences in the properties of the Group 1 elements and the transition elements.

Include at least **two** word equations, or balanced symbol equations, to illustrate your answer.

In this question you will get marks for using good English, organising information clearly, and using scientific words correctly.

....

....

....

....

....

....

....

....

....

....

(6 marks)

(Total marks: 6)

3 The table gives some properties of the Group 7 elements.

Element	Melting point (°C)	Boiling point (°C)	Reaction with hydrogen
fluorine	−220	−118	On mixing the two elements explode. Hydrogen fluoride is formed.
chlorine	−101	−35	A mixture of the two elements explodes when a camera flashes. Hydrogen chloride is formed.
bromine	−7	59	A mixture of the two elements burns quickly when ignited by a lighted splint.

a Use the data in the table to give the state of fluorine at room temperature (20 °C)

....

(1 mark)

b Describe the trend in boiling point as the group is descended from top to bottom.

....

(1 mark)

c Write a word equation for the reaction of hydrogen with fluorine.

Select information from the table to help you.

....

(1 mark)

d i Use the periodic table to identify the element below bromine in Group 7.

....

(1 mark)

ii Select data from the table above to predict how this element reacts with hydrogen.

Include the name of the product and suggest how vigorously the two elements will react.

....

....

(2 marks)

e i Describe the trend in reactivity in the Group 7 elements.

Illustrate your answer by referring to:
- reactions shown in the table at the start of question 3
- the displacement reactions of the Group 7 elements.

....

....

....

....

(4 marks)

H **ii** Explain the trend in reactivity you have described in part **i** above.

....

....

(2 marks)

(Total marks: 12)

Revision objectives

- describe how tap water is made safe to drink
- describe how filters improve water taste and quality
- evaluate the advantages and disadvantages of adding chlorine and fluoride compounds to drinking water
- explain why purifying water by distillation is expensive

Student book references

3.10 Fluorides and filters

3.11 Drinking seawater

Specification key

✔ C3.2.2

Making water safe to drink

Water of the correct quality is vital for life. Microbes and dissolved salts may harm human health.

In the UK, water companies provide safe drinking water by:

- ***Choosing a suitable source:*** Water from different sources needs different treatments. Water from boreholes may have been filtered by underground rock for many years. It would then need little treatment. Canal water may contain algae, bacteria, and fertiliser chemicals. It needs careful treatment.
- ***Filtration:*** Water trickles through huge beds of sand, called **filter beds**. These remove solid impurities.
- ***Sterilisation:*** Chlorine **sterilises** water by killing microbes. The chlorine remains in the water. It kills any microbes that get into the water between the treatment works and people's homes.

▲ The flow diagram summarises the stages by which water is made safe to drink.

Adding fluorides

Some water companies add fluoride compounds to tap water. Fluorides help to prevent tooth decay.

There are arguments for and against adding fluoride compounds to water.

Arguments for adding fluoride to drinking water

- It prevents tooth decay, meaning fewer people suffer from toothache and other dental problems.
- Less money is spent treating dental problems.

Arguments against adding fluoride to drinking water

- It is expensive.
- If everyone looked after their teeth properly there would be little need to add fluorides to water.
- Swallowing very large amounts of fluoride compounds may make teeth go yellow.

Home water filters

Some people use water filters to improve the taste or quality of their tap water.

Ion exchange resin filters

Some types of water filter contain an **ion exchange resin**. Ion exchange resins remove metal ions (for example, lead, copper or cadmium) from the water. The filter replaces these metal ions with less harmful ions, such as hydrogen, sodium, or potassium.

Ion exchange resins are also used to soften hard water.

Carbon filters

Carbon water filters remove chlorine from water, as well as other substances with unpleasant smells or tastes. The filters contain **activated carbon**. This has a huge surface area. As tap water passes through the filter, molecules of unwanted substances stick to the surface of the carbon. The process is called **adsorption**.

Silver filters

Some home water filters contains a source of silver ions, Ag^+. These destroy some types of dangerous bacteria.

Distillation

Pure water can be produced by the **distillation** of seawater. The process requires huge energy inputs. This makes it very expensive.

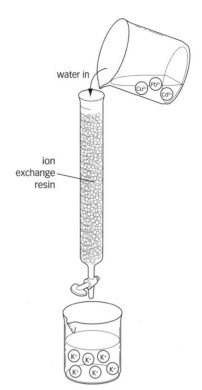

▲ In this water filter, metal ions are replaced by potassium ions. The ions are not drawn to scale.

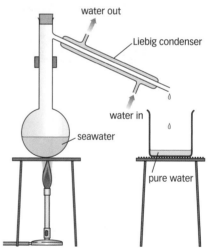

▲ The diagram shows how to produce small amounts of pure water from seawater. Countries such as the United Arab Emirates, which has no rivers or lakes, obtain most of their water by the large-scale distillation of seawater.

Questions

1   List the **three** main stages by which water of the correct quality is provided for British homes.

2   Evaluate the advantages and disadvantages of adding fluoride to water.

3   Draw a table to summarise the substances removed from tap water by these types of water filter – carbon, silver, ion exchange resin.

Key words

filter bed, sterilises, ion exchange resin, activated carbon, adsorption, distillation

Revision objectives

- ✔ explain what makes water hard or soft
- ✔ describe how to measure water hardness
- ✔ evaluate the environmental, social and economic aspects of water hardness
- ✔ describe and explain three water-softening methods

Specification key

✔ C3.2.1

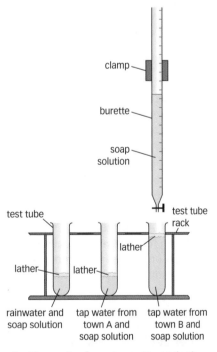

clamp

burette

soap solution

test tube

test tube rack

lather

lather

lather

rainwater and soap solution

tap water from town A and soap solution

tap water from town B and soap solution

▲ The water from town B needs the most soap solution to form lather. This shows that the water from town B is harder than water from town A, and harder than rainwater.

Measuring water hardness

Soft water easily forms lather with soap. **Hard water** reacts with soap to form scum. This means that more soap is needed to form lather with a given volume of hard water than with the same volume of soft water. Soapless detergents never form scum.

You can use soap solution to compare the hardness of different water samples. The more soap solution needed to form a permanent lather (one that doesn't disappear on shaking), the harder the water.

What makes water hard?

Hard water contains dissolved compounds, usually of calcium or magnesium. The compounds are dissolved when water flows through limestone or chalk rock.

Soft water does not contain dissolved calcium or magnesium compounds.

There are two types of hard water:

- **Permanent hard water** remains hard, even when it is boiled. It contains dissolved calcium and sulfate ions.
- **Temporary hard water** is softened by boiling.

H Temporary hard water contains **hydrogen carbonate ions (HCO_3^-)**. On heating, these ions decompose to produce carbonate ions (CO_3^{2-}). The carbonate ions react with calcium ions (Ca^{2+}) or magnesium ions (Mg^{2+}) in the water to make calcium carbonate or magnesium carbonate.

$$\text{calcium hydrogen carbonate} \rightarrow \text{calcium carbonate} + \text{carbon dioxide} + \text{water}$$

$$Ca(HCO_3)_2 \text{ (aq)} \rightarrow CaCO_3 \text{ (s)} + CO_2 \text{ (g)} + H_2O \text{ (l)}$$

Calcium carbonate and magnesium carbonate are insoluble in water, so they form as precipitates. The precipitates form scale in kettles and boilers. The boiled water contains few calcium or magnesium ions. It has been softened.

Exam tip

Don't forget the difference between temporary and permanent hardness – temporary hard water is softened by boiling, permanent hard water remains hard when it is boiled.

Hard and soft water – costs and benefits

Hard water is good for health. Its calcium compounds help with the development and maintenance of teeth and bones.

Calcium compounds also help to reduce heart disease.

Hard water also has disadvantages:
- More soap is needed for washing, since scum is formed. This increases costs.
- Heating temporary hard water in kettles and boilers produces scale. Scale makes kettles and heating systems less efficient. This increases costs and may increase greenhouse gas emissions.

Softening hard water: sodium carbonate

You can soften hard water by adding **sodium carbonate**. Sodium carbonate is soluble in water. In hard water, its carbonate ions react with dissolved calcium and magnesium ions. Calcium carbonate and magnesium carbonate form as precipitates. They can be removed by filtering.

The equations below summarise one of these reactions:

calcium ions + carbonate ions →calcium carbonate

$$Ca^{2+} (aq) + CO_3^{2-} (aq) \rightarrow CaCO_3 (s)$$

Softening hard water: ion exchange

Ion exchange columns swap calcium and magnesium ions with sodium or hydrogen ions.

In the ion exchange column on the right, sodium ions are attached to the resin. Water flows through the column from the top. As the water flows down, calcium and magnesium ions from the water stick to the resin. Sodium ions from the resin dissolve in the water.

After a while the ion exchange column stops working – all its sodium ions have been replaced by calcium and magnesium ions. You then need to pour sodium chloride solution through the column. Sodium ions from the solution stick to the resin. Calcium and magnesium ions are flushed away. The column is ready to use again.

> ### Key words
> soft water, hard water, hydrogen carbonate ion (HCO_3^-), sodium carbonate, ion exchange column

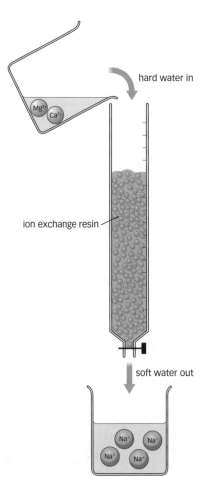

▲ This ion exchange column softens hard water. Ions are not drawn to scale.

Questions

1 Explain what makes water hard.

2 Identify the advantages and disadvantages of hard and soft water.

3 List **three** ways of softening hard water.

133

Working to Grade E

1 In the list below, write **S** next to the statements that are true for soft water. Write **H** next to the statements that are true for hard water.
  a It contains compounds that help bones develop.
  b It needs only a little soap to make lather.
  c When heated it may produce scale in kettles.
  d It helps reduce heart disease.
  e It contains compounds that help maintain healthy teeth.

2 The table shows three steps taken by water companies to produce drinking water. Draw lines to match each step to one reason.

Step	Reason
Choose an appropriate source	to remove solids from the water.
Pass the water through filter beds	to kill bacteria in the water.
Add chlorine	to minimise the treatment needed to make the water safe to drink.

3 Use the words in the box below to complete the sentences that follow. Each word may be used once, more than once, or not at all.

> scale detergents less lather more scum

Soft water easily forms _____ with soap. Hard water reacts with soap to form _____ so _____ soap is needed to form lather. Soapless _____ do not form scum.

4 Put ticks next to the compounds that may be dissolved in hard water to make the water hard.
  a calcium sulfate
  b sodium sulfate
  c calcium hydrogen carbonate
  d magnesium carbonate

Working to Grade C

5 In some areas, water companies add fluoride compounds to water. Identify **two** arguments for adding fluorides to water, and **two** arguments against adding fluorides to water.

6 Describe **one** difference between permanent hard water and temporary hard water.

7 Martha investigates the hardness of water from three villages. She takes three 10 cm³ samples of water from each village. She measures the number of drops of soap solution required to form permanent lather. Her results are in the table below.

Village	Number of drops of soap solution needed to make permanent lather			
	Run 1	Run 2	Run 3	Mean
A	2	3	4	
B	23	17	20	
C	43	45	29	

  a Draw a ring around the anomalous result in the table.
  b Calculate the missing means and write them in the table. Ignore the anomalous result.
  c Suggest why Martha tested three samples of water from each village.
  d Write the letters of the villages in order of increasing water hardness.
  e In which village would you expect little scale to be formed in kettles?
  f Which village or villages are most likely to obtain their water from an underground source in a limestone area?

8 This diagram shows how an ion exchange column makes hard water soft. Write the letters of the labels below in the correct boxes. You may write one or more letters in each box.

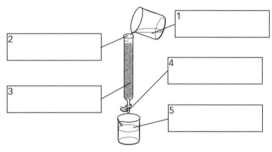

  A Hard water enters the ion exchange column here.
  B After use, calcium or magnesium ions are attached to this.
  C Soft water leaves the ion exchange column here.
  D This water contains dissolved calcium or magnesium ions.
  E This water contains dissolved sodium ions.
  F Before use, sodium ions are attached to this.

9 Explain how adding sodium carbonate softens hard water.

10 Pure water can be produced from seawater by distillation. Explain why, in the UK, the process is very expensive.

Working to Grade A*

11 The statements below describe why temporary hard water forms scale in kettles.
The statements are in the wrong order. Write the letters of the statements in the best order in the boxes at the end of this question.
  A These ions decompose on heating.
  B These deposit in kettles as scale.
  C They react with calcium and magnesium ions.
  D Temporary hard water contains hydrogen carbonate ions.
  E Carbonate ions are produced.
  F Precipitates are formed.

1 Some friends are discussing the advantages and disadvantages of adding chlorine to drinking water.

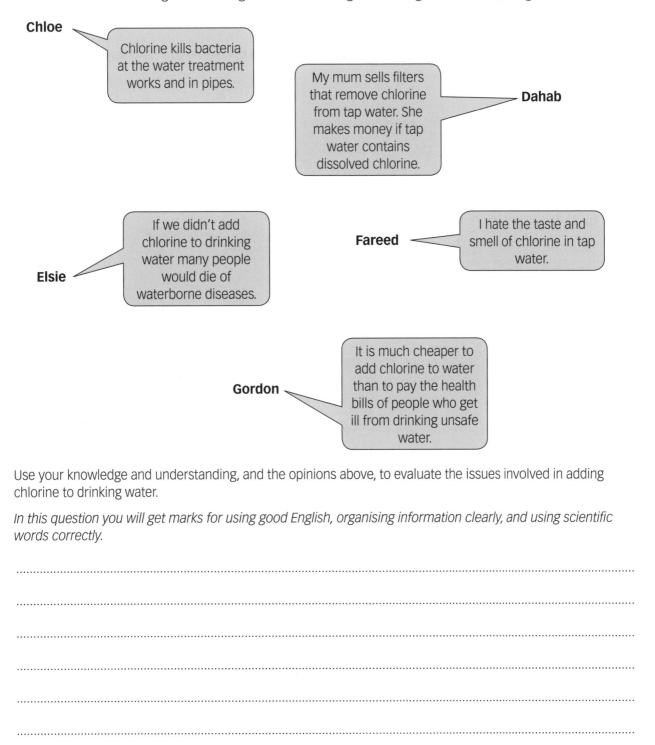

Chloe
Chlorine kills bacteria at the water treatment works and in pipes.

Dahab
My mum sells filters that remove chlorine from tap water. She makes money if tap water contains dissolved chlorine.

Elsie
If we didn't add chlorine to drinking water many people would die of waterborne diseases.

Fareed
I hate the taste and smell of chlorine in tap water.

Gordon
It is much cheaper to add chlorine to water than to pay the health bills of people who get ill from drinking unsafe water.

Use your knowledge and understanding, and the opinions above, to evaluate the issues involved in adding chlorine to drinking water.

In this question you will get marks for using good English, organising information clearly, and using scientific words correctly.

....

....

....

....

....

....

....

....

....

(6 marks)
(Total marks: 6)

2 A student wants to compare the hardness of water in his home on different days.

a Why might the hardness of the water be different on different days?

Tick the **two** best answers.

On different days, the water company adds different amounts of chlorine to the water. ☐

The student has a water softener at home, but it doesn't work when the column is saturated with calcium ions. ☐

On different days, the water company supplies water from different sources. ☐

On warmer days the water dissolves more metal ions as it flows through the pipes. ☐

(2 marks)

b The student uses the apparatus below to test water samples collected on different days.

Describe how the student could use the apparatus to compare the hardness of water samples collected on different days.

....

....

....

....

....

clamp

soap solution

burette

test tube

test tube rack

(3 marks)

c The student's results are in the table below.

Date	Number of drops of soap solution needed to make permanent lather
20 March	10
3 April	2
23 August	3
2 October	59

i From the results in the table, identify the date on which the hardest water sample was collected.

....

(1 mark)

ii Suggest how the student could improve his investigation.

....

(1 mark)

iii On which of the dates in the table would the student expect the most scale to be formed in his kettle?

....

(1 mark)

(Total marks: 8)

3 The diagram shows an ion exchange column used to soften hard water. Before use, sodium ions are attached to the ion exchange resin.

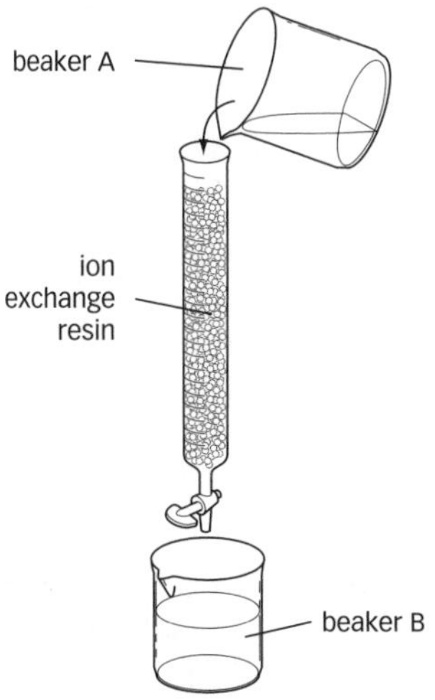

beaker A

ion exchange resin

beaker B

  a **i** Name the **two** ions that may make water hard.

.... and ....

(2 marks)

      ii Compare the water in beaker B with that in beaker A.

....

....

....

(2 marks)

  b Evaluate the advantages and disadvantages of using a home water softener.

....

....

....

(3 marks)

(Total marks: 7)

4 This question is about temporary and permanent hard water.

  a Describe **one** difference between temporary hard water and permanent hard water.

....

(1 mark)

  b Adding sodium carbonate softens both temporary and permanent hard water. Explain how.

....

....

(2 marks)

H

  c Describe and explain one method of softening temporary hard water that does not involve adding a chemical to the water, and does not involve using an ion exchange resin.

....

....

....

....

(4 marks)

(Total marks: 7)

Revision objectives

✔ work out the energy released on burning different fuels

✔ calculate the energy released by reactions in solution

Student book references

3.12 Measuring food and fuel energy

3.13 Energy changes

Specification key

✔ C3.3.1 a – c

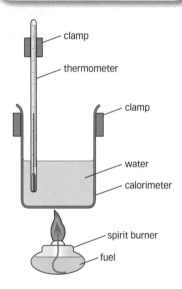

clamp
thermometer
clamp
water
calorimeter
spirit burner
fuel

Exam tip AQA

When calculating the energy released on burning fuels, remember that the 'm' in the equation $Q = mc\Delta T$ is the mass of the water, not the mass of fuel that was burnt.

Measuring energy changes when fuels burn

Burning reactions are **exothermic**. So when fuels burn, energy is released (given out). Different fuels release different amounts of energy. Energy is measured in **joules** (J). One thousand joules is one **kilojoule**, kJ.

You can use the apparatus below to compare the energy released when different fuels burn. The technique is called **calorimetry**.

To find the energy transferred to the water by a burning fuel, follow the steps below:
- Find the total mass of the spirit burner + fuel.
- Measure out a known mass of water.
- Record the water temperature.
- Burn the fuel so it heats the water.
- Record the new water temperature.
- Find the new mass of the spirit burner + fuel.
- Use the equation below to calculate the amount of energy transferred to the water:

$$\frac{\text{energy}}{\text{(in J)}} = \frac{\text{mass of water}}{\text{(in g)}} \times \frac{\text{specific heat}}{\substack{\text{capacity of water} \\ \text{(in J/g°C)}}} \times \frac{\text{temperature}}{\substack{\text{change} \\ \text{(in °C)}}}$$

$$Q = mc\Delta T$$

- Divide the value of Q by the mass of fuel burnt. This is the energy released by 1 g of fuel.

Not all the energy released on burning is transferred to the water. Some is transferred to the calorimeter, some to the rest of the apparatus, and some to the air.

Worked example

George uses two fuels – methanol and ethanol – to heat 100 g of water. He collects the data in the table.

The specific heat capacity of water = 4.2 J/g °C

How much energy does each fuel transfer to the water?

	Methanol	Ethanol
mass of spirit burner + fuel before heating (g)	260.0	270.0
mass of spirit burner + fuel after heating (g)	259.0	269.5
mass of fuel burnt (g)	1.0	0.5
mass of water (g)	100	100
temperature of water before heating (°C)	20	21
temperature of water after heating (°C)	72	50
increase in water temperature (°C)	52	29

For 1g of methanol, energy transferred = $m \times c \times \Delta T$

$= 100 \times 4.2 \times 52$

$= 21\,840\,J$

For 0.5g of ethanol energy transferred = $m \times c \times \Delta T$

$= 100 \times 4.2 \times 29$

$= 12\,180\,J$

So for 1g of ethanol, energy transferred = $12\,180 \div 0.5$

$= 24\,360\,J$

So ethanol transferred more energy to the water per gram of fuel.

Measuring energy changes when foods burn

You can use the apparatus on the right to measure the energy released when foods burn.

Measuring energy changes for reactions in solution

You can calculate the energy released or absorbed by a chemical reaction in solution by measuring the temperature change.

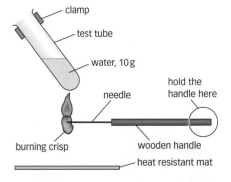

Worked example

Zoë wants to measure the energy transfer for the reaction of a metal powder with a dilute acid. She sets up the apparatus below. She then:

- measures the temperature of the acid
- adds the metal powder, with stirring
- records the maximum temperature reached.

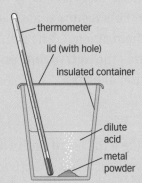

Zoë collects the data in the table.

volume of acid (g)	100
temperature at start (°C)	19
highest temperature reached (°C)	71
temperature change (°C)	52

She uses the equation $Q = mc\,\Delta T$ to calculate the energy transfer. She assumes that it is only the water in the solution that is being heated.

$Q = m \times c \times \Delta T$

$Q = 100 \times 4.2 \times 52$

$Q = 2184\,J$

So the energy change for the reaction is −2184 J. The negative sign shows that the reaction is exothermic.

The same method can be used to measure the energy transferred in any reactions of solids with water or solutions, and in neutralisation reactions.

Questions

1   Give the name and symbol of the unit for energy.

2   Calculate the energy transferred when 1g of a fuel increases the temperature of 100g of water by 70 °C.
    The specific heat capacity of water, c, has a value of 4.2 J/g °C.

3   When measuring the energy transferred from a burning fuel to a container of water, why is it important for the container to be insulated?

Revision objectives

- ✔ use and interpret energy-level diagrams
- ✔ explain energy changes in terms of making and breaking bonds

Student book references

3.14 Energy-level diagrams

3.15 Bond breaking, bond making

Specification key

✔ C3.3.1 d – h

Energy-level diagrams

Energy-level diagrams show the relative energies of reactants and products, and the overall energy change of a reaction.

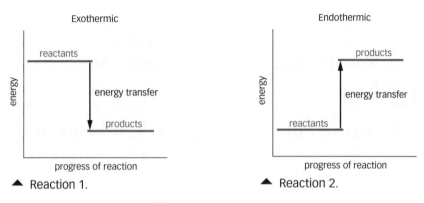

▲ Reaction 1.　　　　　▲ Reaction 2.

The energy-level diagram for reaction 1 shows that the energy stored in the products is less than the energy stored in the reactants. The reaction is exothermic.

The diagram for reaction 2 shows that the reaction is endothermic. The products store more energy than the reactants. Energy changes for endothermic reactions have positive values.

Activation energy

Reactions can only happen when reactant particles collide. Only particles with enough energy can actually react when they do collide. This means that chemical reactions need energy to get them started. The minimum energy needed to start a reaction is the **activation energy**. Every reaction has its own activation energy.

Catalysts

Catalysts provide a different pathway for a chemical reaction. The new pathway has a lower activation energy.

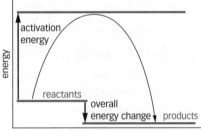

▲ This energy-level diagram shows the overall energy change, and the activation energy, for a reaction. The curved arrow shows the energy as the reaction proceeds.

Exam tip　　AQA

In energy-level diagrams:
- if the reactants are higher than the products, the reaction is exothermic
- if the reactants are lower than the products, the reaction is endothermic.

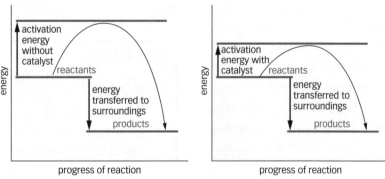

▲ The two energy-level diagrams are for the same reaction. The diagram on the right shows that the activation energy is lower when a catalyst is used.

Bond breaking and bond making

In a chemical reaction:
- energy must be supplied to break bonds – endothermic
- energy is released when new bonds are formed – exothermic.

For example: hydrogen + chlorine→hydrogen chloride

$$H_2 \quad + \quad Cl_2 \quad \rightarrow \quad 2HCl$$

In this reaction, energy must be supplied to break H–H bonds in hydrogen molecules and to break Cl–Cl bonds in chlorine molecules. Energy is released when a hydrogen atom joins with a chlorine atom to make the bond in a hydrogen chloride molecule, H–Cl.

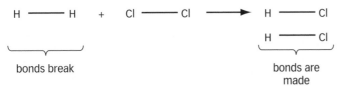

H Endothermic or exothermic?

The difference between the energy needed to break bonds in reactants and the energy released on making new bonds in products determines whether a reaction is exothermic or endothermic.

Every bond has its own **bond energy**, the energy needed to break it. You can use bond energies to calculate energy changes in reactions.

Bond	Bond energy (kJ/mol)
H–H	436
F–F	158
H–F	562

Worked example

Calculate the energy change for the reaction

$$H_2\ (g) + F_2\ (g) \rightarrow 2HF\ (g)$$

The energy needed to break one mole of H–H bonds and one mole of F–F bonds is (436 + 158) = 594 KJ

Two moles of HF form in the reaction. The energy released when its bonds form is (2 × 562) = 1124 kJ

Overall energy transfer
= energy supplied to break bonds – energy released on
          making bonds

= 594 – 1124 = –530 KJ

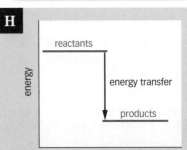

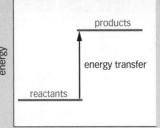

▲ The energy released on forming new bonds is greater than the energy needed to break existing bonds. The reaction is exothermic.

▲ The energy needed to break existing bonds is greater than the energy released from forming new bonds. The reaction is endothermic.

Questions

1   Sketch an energy-level diagram for an exothermic reaction. Label the axes, and the reactants and products.

2   Explain how a catalyst increases the rate of a reaction.

3   H Use ideas about bond energies to explain why the combustion reaction of hydrogen is exothermic. The equation for the reaction is
$$2H_2 + O_2 \rightarrow 2H_2O$$

Revision objectives

✔ describe two ways in which hydrogen is used as a fuel
✔ compare the consequences of using different fuels

Student book references

3.16 Hydrogen fuels

Specification key

✔ C3.3.1 i

Hydrocarbon fuels – pros and cons

Most cars are fuelled by petrol or diesel. These fuels are hydrocarbons. They are obtained from crude oil. They burn in car engines to supply the energy needed to make cars move.

Using hydrocarbon fuels has many consequences.

Category	Consequences
Social	• Travel by car can be convenient. • Increasing car use may make towns less pleasant for pedestrians and cyclists. • Supplies of crude oil will one day run out – we cannot rely on petrol and diesel for ever.
Economic	• Oil companies make money by selling fuels. • When fuel prices increase people find them less affordable.
Environmental	• Burning hydrocarbons makes carbon dioxide, which causes climate change. • Burning diesel makes particulates, which may lead to asthma, lung cancer, and heart disease. • Burning hydrocarbons at high temperature makes oxides of nitrogen. These destroy ozone in the upper atmosphere. Ozone protects us from cancer-causing ultraviolet radiation.

Hydrogen fuel

Engineers are developing hydrogen-fuelled cars. The hydrogen can be:

• burnt in combustion engines

$$\text{hydrogen} + \text{oxygen} \rightarrow \text{water}$$
$$2H_2\,(g) \ + \ O_2\,(g) \rightarrow 2H_2O\,(l)$$

• used to generate electricity in fuel cells.

When hydrogen-powered vehicles are moving, they do not produce carbon dioxide. However, the hydrogen must be manufactured. It is often made by reacting methane with water. The process produces carbon dioxide, a greenhouse gas.

The table shows some advantages and disadvantages of hydrogen fuel cells compared to burning hydrogen in an internal combustion engine.

Exam tip

If you are asked to evaluate the consequences of burning different fuels, try grouping the consequences into categories, for example, environmental, social, and economic.

Questions

1 Describe **two** ways in which hydrogen is used as a fuel.

2 Identify **three** consequences of burning hydrocarbon fuels in cars.

Fuel cells	Hydrogen-fuelled internal combustion engines (ICE)
more efficient than ICEs	less efficient than fuel cells
batteries expensive	technology well understood
include expensive platinum catalysts	nitrogen and oxygen react in the engine to produce nitrogen oxides as well as water
few fuel stations supply hydrogen gas for refuelling	

1 Write **T** next to the statements that are true. Write corrected versions of the statements that are false.

   a Energy is normally measured in joules.

   b 100 J = 1 KJ

   c In the equation $Q = mc\Delta T$, ΔT represents energy change.

   d In an energy-level diagram, a curved arrow shows the energy as the reaction proceeds.

   e In a chemical reaction, energy is released when bonds break.

   f In a chemical reaction, energy must be supplied to form new bonds.

   g A catalyst provides a different pathway for a reaction. The new pathway has a higher activation energy.

2 Look at the energy-level diagram.

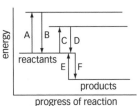

   a Give the letter of the arrow that shows the overall energy change of the reaction.

   b Give the letter of the arrow that shows the activation energy for the reaction without a catalyst.

   c Give the letter of the arrow that shows the activation energy for the reaction when a catalyst is used.

   d Is the reaction exothermic or endothermic? Explain how you decided.

3 Use data from the table to help you answer the questions below.

Fuel	State at room temperature	Energy released on burning fuel (kJ/g)
methane	gas	56
ethanol	liquid	30
hydrogen	gas	143

   a Per gram, which fuel transfers most energy on burning?

   b Identify some advantages and disadvantages of using hydrogen as a fuel compared to the other fuels in the table.

4 Eduardo uses a burning crisp to heat 100 g of water. The temperature of the water rose by 35 °C. Use the equation $Q = mc\Delta T$ to calculate the energy released by the burning crisp. The specific heat capacity of water is 4.2 J/g°C.

5 Clarissa adds 50 cm³ of sodium hydroxide solution to 50 cm³ of hydrochloric acid solution in an insulated container. The temperature of the water increases from 20 °C to 27 °C.

   a Calculate the energy change for the reaction.

   b Is the reaction exothermic or endothermic? Explain how you decided.

6 Use the data on the energy-level diagrams below to answer the questions that follow.

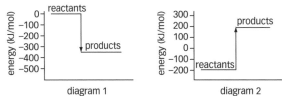

   a Give the overall energy change, including a + or – sign for:

     i the reaction shown in diagram 1

     ii the reaction shown in diagram 2.

   b Which of the reactions shown on the energy-level diagrams releases energy? Explain how you decided.

7 For each reaction below, draw every bond that breaks and every bond that is made in the correct boxes in the table.

   a $H_2\,(g) + Cl_2\,(g) \rightarrow 2HCl\,(g)$

   b $2H_2\,(g) + O_2\,(g) \rightarrow 2H_2O\,(g)$

   c $CH_4\,(g) + Cl_2\,(g) \rightarrow CH_3Cl\,(g) + HCl\,(g)$

Reaction	Bonds that break	Bonds that are made
a		
b		
c		

8 The table shows the energy needed to break some covalent bonds. Use data from the table to answer the questions below it.

Bond	Bond energy (kJ/mol)	Bond	Bond energy (kJ/mol)
H–H	436	H–Cl	432
O=O	497	C–H	413
H–O	463	C–Cl	339
Cl–Cl	243		

   a Which bond needs most energy to break it?

   b Which bond gives out least energy when it is made?

   c Which bond is strongest?

9 Calculate energy changes for the reactions below. To help you, use your answers to question 7 and the data in question 8.

   a $H_2\,(g) + Cl_2\,(g) \rightarrow 2HCl\,(g)$

   b $2H_2\,(g) + O_2\,(g) \rightarrow 2H_2O\,(g)$

   c $CH_4\,(g) + Cl_2\,(g) \rightarrow CH_3Cl\,(g) + HCl\,(g)$

1 A student does an experiment to compare the energy stored in two types of breakfast cereal, WheetyWheels and RicyRings.

He sets up the apparatus below.

He burns 1 g of WheetyWheels, and uses it to heat 100 g of water.

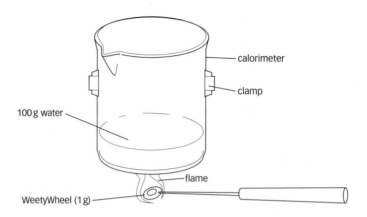

For WheetyWheels, the student obtains the data in the table.

	Run 1	Run 2	Run 3
Initial water temperature (°C)	19	20	22
Final water temperature (°C)	51	56	60
Temperature change (°C)	32	36	38

a Suggest how the student could ensure his investigation was fair.

....

(2 marks)

b Use the data in the table to calculate the mean temperature change for WheetyWheels.

Show clearly how you work out your answer.

....

Mean temperature change = ... °C

(1 mark)

c Calculate how much energy is transferred to the water by 1 g of WheetyWheels. Use the equation

energy released = mass of water × 4.2 × mean temperature change
(in J)   (in g)   (in °C)

Show clearly how you work out your answer.

....

....

Energy released on burning 1 g of WheetyWheels = ... J

(2 marks)

d The student then uses 1 g of burning RicyRings to heat the water.

He calculates that 10000 J is transferred to the water.

He compares his value to the value on the RicyRings packet.

The packet states that burning 1 g of RicyRings releases 16000 J.

i Suggest **one** reason for the difference between the student's value and the value on the RicyRings packet.

....

(1 mark)

ii Suggest **one** change the student could make to his apparatus to improve the data he collects.

....

(1 mark)

(Total marks: 7)

2 A farmer plans to collect the methane gas produced by decaying cow manure.

The methane gas will be piped to local homes and burnt to provide energy for cooking and heating.

Methane is a greenhouse gas.

The equation below shows the products formed when methane burns.

$$\text{methane} + \text{oxygen} \rightarrow \text{carbon dioxide} + \text{water}$$
$$CH_4 + 2O_2 \rightarrow CO_2 + 2H_2O$$

The equation can also be written showing the structural formulae.

a **i** Identify two environmental impacts of the farmer's scheme.

1 ....

2 ....

(2 marks)

ii Suggest an economic benefit of the scheme.

....

(1 mark)

b The combustion (burning) of methane is an exothermic reaction.

Explain how the energy-level diagram below shows that the reaction is exothermic.

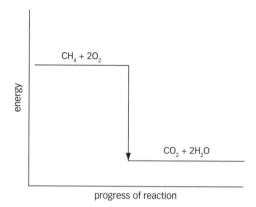

progress of reaction

....

....

....

....

(2 marks)

c Use the bond energies in the table below, and the equations at the start of question 2, to calculate the energy change for the combustion of methane.

Show clearly how you work out your answer.

Bond	Bond energy (kJ/mol)
C=O	743
O=O	497
H–O	463
C–H	413

....

....

....

....

Energy released on burning one mole of methane = ... kJ

(4 marks)

(Total marks: 9)

Presenting data and using data to draw conclusions

In this module there are several opportunities to present data and draw conclusions. These include presenting data and drawing conclusions on water hardness, softening water, and energy changes in chemical reactions.

As well as demonstrating these skills practically, you are likely to be asked to comment on data presented by others, and on the conclusions others draw from data. The examples below offer guidance in these skill areas. They also give you the chance to practise using your skills to answer the sorts of questions that may well come up in exams.

Comparing the hardness of water from different towns

> 1 Irma did tests to compare the hardness of water from five towns. She took $10\,cm^3$ water from each town. She added soap solution to each sample, and counted the number of drops required to make a permanent lather. For each town, she repeated the test three times.
>
> Her results are in the table.

Town	Number of drops of soap solution required for permanent lather			
	Run 1	Run 2	Run 3	Mean
A	2	5	8	
B	28	26	30	
C	6	6	9	
D	58	62	54	
E	6	7	8	

> a Of all the towns in the table, identify the town that has the greatest range for the number of drops of soap solution required.

The range of a data set refers to the maximum and minimum values.
- For town A the range is 2 to 8 (a difference of 6).
- For town B the range is 26 to 30 (a difference of 4).

So of the two towns above, town A has the greater range.

> b Calculate the mean number of drops of soap solution required for each town.

The mean of the data refers to the sum of all the measurements divided by the number of measurements taken.
- For town A the mean = $(2 + 5 + 8) \div 3 = 5$
- For town B the mean = $(28 + 26 + 30) \div 3 = 28$

> c Would it be better to draw a line graph or a bar chart to display the data in the table? Give a reason for your decision.

- If one of the variables is categoric (it can be described by a label, and has no numerical meaning), use a bar chart to display the data.
- If both the dependent and independent variables are continuous (can have any numerical value), use a line graph to display the data.

> d Draw a bar chart or a line graph to display the data in the table.

When drawing a bar chart or line graph, remember the following:
- Label both axes clearly. Don't forget to include units, if they are needed.
- Choose sensible scales.
- Plot the data accurately.

Softening water

Now follow the examiner's advice given for the questions above by answering the exam question below.

> 2 Jake investigated the effect of adding different amounts of sodium carbonate to $10\,cm^3$ samples of hard water from the same source.
>
> A summary of his results is in the table.

Volume of sodium carbonate solution added (cm^3)	Number of drops of soap solution required for permanent lather
1	38
2	29
3	21
4	10
5	2
6	2
7	15
8	2

> a Draw a line graph to display the data in the table.
> b Use your graph to identify any anomalous values in the data set.

An anomalous value is one that does not fit the pattern shown by the rest of the data. Displaying data on a line graph may help you to spot anomalous data.

> c Write a sentence or two to describe the relationship shown on your graph.

AQA Upgrade

Answering an extended writing question

1 The two energy-level diagrams below are for the same reaction. Explain and compare what the two diagrams show, suggesting reasons for any differences. In your answer, refer to the labelled arrows. *(6 marks)*

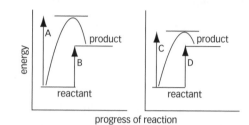

QUESTION

G–E

they both have the same overall energie change and the arrows of some are the same and some are different and there is a curved lighn and some strayt lighns but i dont no why.

Examiner: This answer is typical of a grade-G candidate. It is worth just one mark, gained for recognising that the overall energy change is the same in both diagrams.

The candidate has used one specialist term. There is no punctuation, and there are several spelling mistakes.

D–C

The reactions is endothermic. I knows this becuase the products have more energy than the reactants (arrow B). Arrows C and A are different lenghs this is to do with activation energy – in diagram C the activation energy is less.
Energy is measured in joules and hydrogen can be burnt as a fuel in combustion engines.

Examiner: This answer is worth three marks out of six. It is typical of a grade-C or -D candidate.

The candidate has correctly pointed out that the reaction is endothermic, and referred to three of the four arrows. The candidate has recognised that arrows C and A refer to activation energy, but has not suggested a reason for the different activation energy values.

The answer is well organised, with a few mistakes of grammar and punctuation. There are two spelling mistakes. The last part of the answer is not relevant.

B–A*

Both diagrams show that the relative energy of the products is greater than that of the reactants. So the reaction is endothermic.
Arrows B and D show the same difference in energy. They represent the overall energy change of the reaction. Since both diagrams represent the same reaction, the overall energy change shown on both diagrams is the same.
Arrows A and C represent activation energy. The activation energy on diagram 1 (arrow A) is greater than that on diagram 2 (arrow C). This could be because a catalyst has been used in the reaction represented by diagram 2. Catalysts provide a different pathway for a reaction that has a lower activation energy.

Examiner: This is a high-quality answer, typical of an A* candidate. It is worth six marks out of six.

The candidate has explained and compared the features of the energy-level diagrams very clearly, and referred to the labelled arrows. The reason given for the difference in the two diagrams is correct, and explained in detail.

The answer is well organised. The spelling, punctuation, and grammar are faultless. The candidate has used several specialist terms.

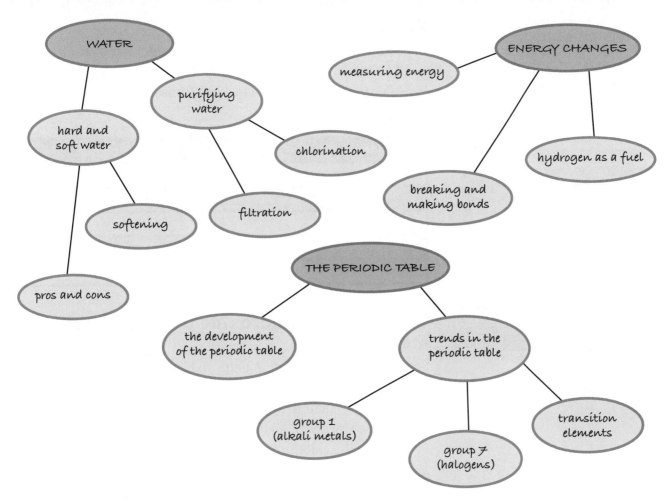

Revision checklist

- Newlands and Mendeleev classified the elements by arranging them in atomic weight order.
- Mendeleev overcame problems in his periodic table by leaving gaps for undiscovered elements.
- In the modern periodic table, elements are arranged in atomic number order.
- Group 1 elements are low-density metals that react with non-metals to form ionic compounds, and with water to release hydrogen. Their hydroxides form alkaline solutions.
- Compared to the Group 1 elements, the transition elements are stronger, harder, have higher melting points, and are less reactive. They have ions with different charges, form coloured compounds, and are useful as catalysts.
- The Group 7 elements react with metals to form ionic compounds. More reactive halogens displace less reactive halogens from aqueous solutions of their salts.
- Hard water contains dissolved compounds of calcium or magnesium. Calcium compounds are good for heart health, teeth, and bones.
- Hard water increases costs because more soap is needed, and scale makes kettles and heating systems less efficient.

- Temporary hard water is softened by boiling. Permanent hard water remains hard when it is boiled.
- Hard water can be softened by adding sodium carbonate or by passing it through ion exchange columns.
- Water is made safe to drink by filtering, and by adding chlorine to reduce microbes. Adding fluoride may improve dental health.
- The energy change of a chemical reaction can be calculated using the equation $Q = mc\Delta T$. ΔT is the temperature change of water heated by a burning fuel, or of reacting solutions.
- Energy-level diagrams show the relative energies of reactants and products, the activation energy, and the overall energy change of a reaction.
- In chemical reactions, energy is supplied to break bonds, and energy is released when new bonds are formed.
- Catalysts reduce the minimum amount of energy needed to start a chemical reaction (the activation energy).
- Hydrogen can be burnt as a fuel in combustion engines, or used in fuel cells to produce electricity to power vehicles.

Flame tests

You can use a flame test to identify the metal ion in a salt.

- Dip the end of a clean nichrome wire in the salt.
- Hold the end of the wire in a hot Bunsen flame.
- Observe the flame colour.

Metal ion	Flame colour
lithium	crimson
sodium	yellow
potassium	lilac
calcium	red
barium	green

Using sodium hydroxide to identify other metal ions

Sodium hydroxide solution helps to identify many other metal ions.

- Dissolve the salt in pure water.
- Add a few drops of sodium hydroxide solution.
- Observe the colour of any precipitate formed.

Metal ion	Metal ion formula	Colour of hydroxide precipitate
aluminium	Al^{3+}	white – dissolves in excess sodium hydroxide solution to form a colourless solution
calcium	Ca^{2+}	white
magnesium	Mg^{2+}	white
copper(ii)	Cu^{2+}	blue
iron(ii)	Fe^{2+}	green
iron(iii)	Fe^{3+}	brown

Questions

1. Give the colours of these precipitates: copper hydroxide, magnesium hydroxide, and iron(ii) hydroxide.

2. Describe the test for aluminium ions, and the results you would expect.

3. Describe the test for carbonates, and the results you would expect.

Testing for carbonates

- Add a few drops of dilute hydrochloric acid to the solid.
- If it fizzes, a gas is being made.
- Test the gas with limewater. A white precipitate may form, which makes the limewater cloudy. This shows that the gas is carbon dioxide, and the solid is a carbonate.

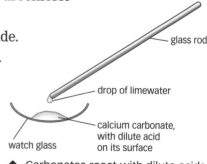

glass rod

drop of limewater

calcium carbonate, with dilute acid on its surface

watch glass

▲ Carbonates react with dilute acids to form carbon dioxide gas.

Testing for halide ions

Use silver nitrate solution to test for chloride, bromide, and iodide ions.

- Dissolve a little of the solid in dilute nitric acid.
- Add silver nitrate solution.

Halide ion	Colour of precipitate	Formula of precipitate
chloride	white	AgCl
bromide	cream	AgBr
iodide	yellow	AgI

The equation for the reaction of sodium bromide with silver nitrate is:

sodium bromide + silver nitrate → silver bromide + sodium nitrate

The ionic equation for the reaction is:

$$Br^- (aq) + Ag^+ (aq) \rightarrow AgBr (s)$$

Testing for sulfate ions

To find out if a solid or solution contains sulfate ions, SO_4^{2-}, follow the steps below.

- Dissolve the solid in dilute hydrochloric acid.
- Add barium chloride solution.

If a white precipitate forms, the sample includes a sulfate ion.

The ionic equation for the reaction is:

$$Ba^{2+} (aq) + SO_4^{2-} (aq) \rightarrow BaSO_4 (s)$$

Finding the volumes of acids and alkalis that react

Titrations show the volumes of acids and alkali solutions that react.

To find the volume of hydrochloric acid that exactly reacts with 25.00 cm³ of sodium hydroxide solution, follow the steps below.

1 Use a pipette to measure out 25.00 cm³ sodium hydroxide solution.
2 Transfer the solution to a conical flask.
3 Add a few drops of indicator solution to the sodium hydroxide solution in the conical flask. Phenolphthalein indicator will turn pink.
4 Pour hydrochloric acid into a burette. Read the scale.
5 Place the conical flask under the burette. Allow the hydrochloric acid to run into the conical flask. Swirl the mixture in the flask.

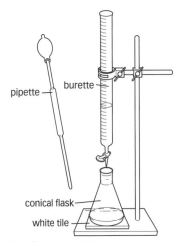

▲ Titration apparatus.

Exam tip

Repeating a titration several times, and calculating an average volume, helps to ensure an accurate result (one that is close to the true value).

6 When the phenolphthalein indicator just turns colourless, the titration has reached its **end point**.
7 Read the scale on the burette. Calculate the volume of acid added. You have now finished the **rough titration**.
8 Repeat steps 1–7. This time, add the acid one drop at a time as you approach the end point. Swirl after every drop. Repeat until you have three consistent values for the acid volume.

Calculating titration volumes

The table gives titration results for neutralising 25.00 cm³ of sodium hydroxide with 2.00 mol/dm³ hydrochloric acid from a burette.

	Rough	Run 1	Run 2	Run 3
initial burette reading (cm³)	0.90	13.10	25.90	0.10
final burette reading (cm³)	14.00	25.90	38.60	12.90
volume of acid added (cm³)	13.10	12.80	12.70	12.80

You can use the results in the table to calculate the mean volume of acid added. Do not include the value for the rough titration.

Mean volume = (12.80 + 12.70 + 12.80) ÷ 3 = 12.77 cm³

H **Worked example**

Calculate the concentration of a solution that has 0.25 moles of sodium hydroxide in 250 cm³ of solution.

Volume of solution in dm³
$$= \frac{250\,cm^3}{1000\,cm^3}$$
$$= 0.25\,dm^3$$

Concentration in mol/dm³
$$= \frac{number\ of\ moles}{volume\ in\ dm^3}$$
$$= \frac{0.25\,mol}{0.25\,dm^3}$$
$$= 1\,mol/dm^3$$

H **Calculating concentration**

Concentration is the amount of solute per unit volume of solution.

A **solute** is a substance dissolved in solution.

Chemists measure concentration in grams per cubic decimetre (g/dm³) or in moles per cubic decimetre (mol/dm³).

One mole (mol) of a substance is its unit mass in grams.

$$\text{concentration (g/dm}^3) = \frac{\text{mass of solute (g)}}{\text{volume of solution (dm}^3)}$$

or

$$\text{concentration (mol/dm}^3) = \frac{\text{number of moles of solute (mol)}}{\text{volume of solution (dm}^3)}$$

Questions

1 Describe the test for bromide ions, and the result you would expect.

2 Explain why a titration should be carried out several times, until consistent results are obtained.

3 **H** Calculate the concentration of a solution that has 0.10 mole of acid dissolved in 500 cm³ of solution.

Calculating masses and moles

If you know the concentration of a solution, and its volume, you can calculate the mass of solute, or the number of moles of solute, in a given volume of solution.

Worked example

What mass of potassium nitrate is in $500\,cm^3$ of a $202\,g/dm^3$ solution?

Volume of solution in $dm^3 = \dfrac{500\,cm^3}{1000\,cm^3} = 0.5\,dm^3$

Mass = concentration in g/dm^3 × volume in dm^3

$= 202\,g/dm^3 \times 0.5\,dm^3 = 101\,g$

Using titrations to calculate concentrations

If you know the concentration of one reactant, you can use titration results to work out the concentration of the other reactant.

Worked example

Rosie has a solution of sodium hydroxide of unknown concentration. She measures $25.00\,cm^3$ of the sodium hydroxide solution into a conical flask. She titrates the sodium hydroxide solution with sulfuric acid of concentration $0.500\,mol/dm^3$. The volume of sulfuric acid required is $24.5\,cm^3$. Calculate the concentration of the sodium hydroxide solution.

Calculate the number of moles of sulfuric acid.

Number of moles = concentration in mol/dm^3 × volume in dm^3

$= 0.500\,mol/dm^3 \times (24.5 \div 1000)\,dm^3$

$= 0.0123\,mol$

Write a balanced equation for the reaction. Use it to work out the number of moles of sodium hydroxide in $25.00\,cm^3$ of solution.

$$2NaOH(aq) + H_2SO_4\,(aq) \rightarrow Na_2SO_4\,(aq) + 2H_2O\,(l)$$

The equation shows that 2 moles of sodium hydroxide react with 1 mole of sulfuric acid.

Rosie has $0.0123\,mol$ of sulfuric acid.

So the number of moles of sodium hydroxide = (0.0123×2)

$= 0.0246$

Calculate the concentration of sodium hydroxide in mol/dm^3

concentration $= \dfrac{\text{number of moles}}{\text{volume in } dm^3} = \dfrac{0.0246\ \text{mol}}{(25 \div 1000)\,dm^3}$

$= 0.984\,mol/dm^3$

Revision objectives

✓ use titration results to calculate concentrations of solutions

Student book references

3.21 Titrations – 2

3.22 Titrations – 3

Specification key

✓ C3.4.1 h

Exam tip AQA

Practise doing as many titration calculations as possible.

Questions

1 Calculate the number of moles of hydrochloric acid in $25.00\,cm^3$ of $2.00\,mol/dm^3$ solution.

2 Calculate the mass of potassium hydroxide in $24.00\,cm^3$ of $0.2\,mol/dm^3$ solution.

3 Izzy uses $20.7\,cm^3$ of sodium hydroxide solution of concentration $8.0\,g/dm^3$ to neutralise $20.0\,cm^3$ of sulfuric acid. Calculate the concentration of the sulfuric acid.

1 Complete the table below.

Compound of...	Flame colour
lithium	
sodium	
potassium	
calcium	
barium	

2 Draw lines to match each precipitate to its colour.

Name of precipitate		Colour
aluminium hydroxide		blue
copper(ii) hydroxide		white
iron(ii) hydroxide		brown
calcium hydroxide		green
iron(iii) hydroxide		white
magnesium hydroxide		white

3 Draw a ring around the name of the precipitate that dissolves in excess sodium hydroxide solution.

  calcium hydroxide   aluminium hydroxide
        magnesium hydroxide

4 Use the words in the box below to complete the sentences that follow. Each word may be used once, more than once, or not at all.

> hydrochloric white silver nitrate sulfuric
>   yellow green barium chloride cream
>   barium sulfate nitric

To test for sulfate ions in a solution, add _____ acid and _____ _____ solution. If a _____ precipitate forms, sulfate ions are present in the solution.
To test for chloride ions in a solution, add _____ acid and _____ _____ solution. If a _____ precipitate forms, chloride ions are present.

5 Describe how to find out whether a compound is a carbonate. Include the names of the chemicals you would use and describe the observations you would expect to make.

6 Dan writes out the steps for doing a titration to find the volume of hydrochloric acid that reacts with $25.00\,cm^3$ of sodium hydroxide solution. He makes **one** mistake in **each** step. Write a corrected version of each step.
   a Use a measuring cylinder to measure out exactly $25.00\,cm^3$ of sodium hydroxide solution.

   b Transfer the solution to a beaker.
   c Using a funnel, pour the hydrochloric acid into a burette. Add a few drops of indicator to the burette.
   d Read the scale on the burette. Add hydrochloric acid from the burette to the sodium hydroxide solution in the beaker until the indicator changes colour. Read the scale on the burette and calculate the volume of acid added.
   e Repeat steps **a** to **d** once more.

7 Sabrina does a set of three titrations. Her results are in the table.

	Rough	Run 1	Run 2	Run 3
initial burette reading (cm³)	0.20	10.20	9.80	19.50
final burette reading (cm³)	10.20	20.00	19.50	29.30
volume of acid added (cm³)	10.00	9.80		9.80

   a Calculate the volume of acid added in run 2.
   b Calculate the mean volume of acid added for runs 1, 2, and 3.

8 Calculate the concentrations of the solutions in the table. Give your answers in g/dm^3.

Solute		Volume of solution (cm³)
Name	Mass of solute in solution (g)	
copper sulfate	8	2000
sodium chloride	15	500
magnesium sulfate	20	100

9 Calculate the number of moles of sulfuric acid in $25.00\,cm^3$ of a $1.0\,mol/dm^3$ solution.
10 Calculate the number of moles of sodium hydroxide in $47.00\,cm^3$ of a $0.1\,mol/dm^3$ solution.
11 Calculate the mass of sodium chloride in $1.00\,dm^3$ of a $1.0\,mol/dm^3$ solution. The formula of sodium chloride is $NaCl$.
12 Calculate the mass of magnesium chloride in $500.00\,cm^3$ of a $0.50\,mol/dm^3$ solution. The formula of magnesium chloride is $MgCl_2$.
13 Rob uses $23.10\,cm^3$ of hydrochloric acid of concentration $1.00\,mol/dm^3$ to neutralise $25.00\,cm^3$ of sodium hydroxide solution. Calculate the concentration of the sodium hydroxide solution. Give your answer in mol/dm^3.
14 Mel places $25.00\,cm^3$ of nitric acid in a flask. She uses $13.40\,cm^3$ of $0.10\,mol/dm^3$ sodium hydroxide solution to neutralise the acid. Calculate the concentration of the nitric acid. Give your answer in mol/dm^3.

Examination questions
Further analysis and quantitative chemistry

1 A student has a mixture of two white powders, X and Y.

She does a series of tests to identify the substances in the mixture.

Her results are in the table.

Test number	Test	Observations
1	Flame test.	Red or crimson flame – not sure which.
2	Dissolve in pure water. Add sodium hydroxide solution.	White precipitate. Insoluble in excess sodium hydroxide.
3	Add dilute hydrochloric acid to the mixture of powders.	Fizzed. The bubbles made limewater cloudy.
4	Dissolve in pure water. Add sulfuric acid and barium chloride solution.	White precipitate.
5	Dissolve in pure water. Add nitric acid and silver nitrate solution.	White precipitate.

a Describe how to do a flame test.

....

....

(1 mark)

b Suggest **three** alternative conclusions the student might make after tests 1 and 2.

Use evidence from the table to support each possible conclusion.

....

....

....

(3 marks)

c Which test described in the table is not valid? Explain why.

....

(1 mark)

d Name the **two** negative ions that are present in the mixture.

Give reasons for your decision.

....

....

(2 marks)

(Total marks: 7)

2 Describe how to do a titration to find out the volume of hydrochloric acid that reacts with a given volume of sodium hydroxide solution.

Include the names of each piece of apparatus used.

In this question you will get marks for using good English, organising information clearly, and using scientific words correctly.

....

....

....

....

....

....

....

....

....

(6 marks)
(Total marks: 6)

3 A student investigated the claim on a carton of blackcurrant drink.

'Our blackcurrant drink has four times more vitamin C than orange juice.'

a The student first titrated 10.00 cm³ samples of **orange juice** with DCPIP solution.

DCPIP reacts with vitamin C. The more DCPIP required, the greater the amount of vitamin C in the drink sample.

The titration results are in the table.

	Rough	Run 1	Run 2	Run 3
Initial burette reading (cm³)	1.00	13.60	25.50	37.50
Final burette reading (cm³)	13.60	25.50	37.50	49.60
Volume of DCPIP added	12.60			

i Explain why the student did a rough titration.

....

(1 mark)

ii Explain why the student repeated the titration several times.

....

(1 mark)

iii Use the results in the table to show that the mean volume of DCPIP that reacts with $10\,cm^3$ of orange juice is $12.00\,cm^3$.

You may write in the table above.

....

(1 mark)

b The student then titrated $10.00\,cm^3$ samples of the **blackcurrant drink** with DCPIP.

The mean volume of DCPIP required was $0.15\,cm^3$.

Do the results of the titrations with the orange juice and blackcurrant drink support the claim on the blackcurrant drink carton?

Explain your answer.

....

(1 mark)
(Total marks: 4)

H

4 Jordan does a titration. He uses $23.00\,cm^3$ of sulfuric acid of concentration $9.8\,g/dm^3$ to neutralise $25.00\,cm^3$ of sodium hydroxide solution.

Calculate the concentration of the sodium hydroxide solution in mol/dm^3.

You are advised to write an equation and show all your working.

....

....

....

....

....

....

(4 marks)

(Total marks: 4)

Revision objectives

- name the raw materials for the Haber process, and identify their sources
- describe the Haber process
- evaluate the conditions necessary in the Haber process to maximise yield and minimise environmental impact
- evaluate the conditions used in the Haber process in terms of energy requirements

Student book references

3.23 Making ammonia

3.24 Conditions for the Haber process

Specification key

✔ C3.5.1 a – b

Raw materials

Ammonia is a compound of nitrogen and hydrogen. Its formula is NH_3. Compounds made from ammonia are vital fertilisers.

Ammonia is manufactured by the **Haber process**. The raw materials are its elements:

- Nitrogen is separated from the air.
- Hydrogen may be obtained from natural gas. Methane from the natural gas reacts with water. The products are carbon monoxide and hydrogen:

$$\text{methane} + \text{water} \rightarrow \text{carbon monoxide} + \text{hydrogen}$$

Hydrogen for the Haber process is sometimes obtained from other sources.

The Haber process

First, the raw materials for the Haber process are purified. Then the pure nitrogen and hydrogen enter the reaction vessel. In the reaction vessel:

- the gases pass over an iron catalyst
- the temperature is about 450 °C
- the pressure is about 200 atm.

Some of the hydrogen and nitrogen react together to form ammonia.

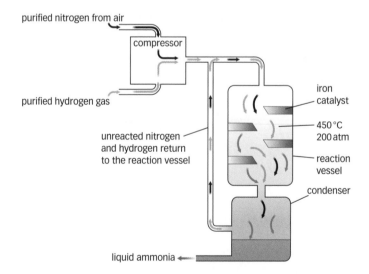

The Haber process reaction is **reversible**. Whilst some ammonia molecules are being made, others are breaking down to form nitrogen and hydrogen. The ⇌ sign shows that the reaction is reversible.

$$\text{nitrogen} + \text{hydrogen} \rightleftharpoons \text{ammonia}$$
$$N_2 \quad + \quad 3H_2 \quad \rightleftharpoons \quad 2NH_3$$

There are three gases in the reaction vessel – nitrogen, hydrogen, and ammonia.

The mixture of gases moves from the reaction vessel to the condenser. In the condenser, the mixture cools. At –33 °C, ammonia gas condenses to form liquid ammonia. The liquid ammonia is removed.

Nitrogen and hydrogen gases remain in the condenser. They are returned to the reaction vessel.

Choosing conditions

Ammonia production companies need to maximise their income. They must make as much ammonia as possible, as quickly as possible. They choose conditions for the Haber process that will maximise:

- the **yield** of ammonia (the percentage of ammonia in the equilibrium mixture)
- the rate of the reaction (the speed at which ammonia is produced).

Pressure

The higher the pressure, the greater the yield of ammonia. But the higher the pressure, the stronger the reaction vessel and pipes need to be, and the more expensive they are. Chemical engineers choose a compromise pressure of about 200 atm – this produces an acceptable yield at a reasonable cost.

Temperature

The lower the temperature, the higher the yield of ammonia. But at low temperatures the reaction is very slow. Chemical engineers choose a compromise temperature of 450 °C. This produces an acceptable yield at a reasonable rate.

Energy and environment

Ammonia companies need to keep their energy costs as low as possible. One way of reducing energy requirements is to transfer the energy released as the reaction mixture cools in the condenser to the reaction vessel.

Reducing energy costs helps to reduce the environmental impact of producing ammonia. There are strict laws to prevent companies allowing poisonous ammonia to escape into the air or water.

Exam tip

Remember the conditions chosen for the Haber process: temperature = 450 °C and pressure = 200 atmospheres (atm).

Questions

1 Name the raw materials for the Haber process, and state the source of each one.

2 Evaluate the temperature chosen for the Haber process in terms of yield, environmental impact, and energy requirements.

3 Evaluate the choice of pressure for the Haber process in terms of yield and cost.

Specification key

✔ C3.5.1 c – h

Equilibrium

Many reactions are reversible. This means they can go in both directions. For example:

- Ammonia gas and hydrogen chloride gas react to form ammonium chloride at room temperature:

    ammonia + hydrogen chloride → ammonium chloride

    $NH_3\,(g)\ +\quad HCl\,(g)\qquad \rightarrow \qquad NH_4Cl\,(g)$

- If you heat ammonium chloride, it decomposes to make ammonia and hydrogen chloride:

    ammonium chloride → ammonia + hydrogen chloride

    $NH_4Cl\,(g)\qquad \rightarrow\ NH_3\,(g)\ +\qquad HCl\,(g)$

In a **closed system**, substances cannot enter or leave the reaction vessel. If a reversible reaction happens in a closed system, it reaches **dynamic equilibrium**. In dynamic equilibrium, the forward and backward reactions are both happening. The rates of both reactions are the same. The amount of each substance in the reaction mixture does not change. For example:

    ammonia + hydrogen chloride ⇌ ammonium chloride

    $NH_3\,(g)\ +\qquad HCl\,(g)\qquad \rightleftharpoons \qquad NH_4Cl\,(g)$

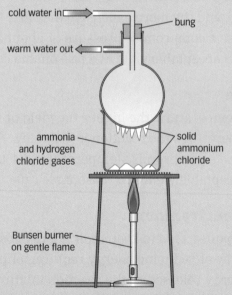

cold water in

bung

warm water out

ammonia and hydrogen chloride gases

solid ammonium chloride

Bunsen burner on gentle flame

▲ The equilibrium of ammonia, hydrogen chloride, and ammonium chloride.

Changing equilibrium conditions

The amounts of substances at equilibrium depend on the conditions, including:

- temperature (for all reactions)
- pressure (for gaseous reactions).

For any equilibrium reaction, if conditions are changed the reaction tends to counteract the effect of the change.

The effect of temperature on equilibrium

For the Haber process reaction below, the forward reaction releases energy in the form of heat. It is exothermic.

$$N_2 + 3H_2 \rightleftharpoons 2NH_3 \quad \Delta H = -92\,KJ/mol$$

If the temperature of the reaction mixture *decreases*, the equilibrium shifts to the right. This releases more energy, and so tends to counteract the effect of decreasing the temperature.

If the temperature of the reaction mixture *increases*, the equilibrium shifts to the left. This absorbs more energy, and so tends to counteract the effect of increasing the temperature. Overall, then:

- If the temperature is raised, the yield from the endothermic reaction increases and the yield from the exothermic reaction decreases.
- If the temperature is lowered, the yield from the endothermic reaction decreases and the yield from the exothermic reaction increases.

The effect of pressure on equilibrium

For gaseous reactions, changing the pressure changes the position of equilibrium. An increase in pressure favours the reaction that produces fewer molecules, as shown by the equation for the reaction.

For example:

$$2SO_2\,(g) + O_2\,(g) \rightleftharpoons 2SO_3\,(g)$$

Number of molecules on left $= (2 + 1) = 3$
Number of molecules on right $= 2$

So increasing the pressure shifts the equilibrium to the right, increasing the yield of sulfur trioxide, SO_3

A second example is the equilibrium reaction of nitrogen, hydrogen, and ammonia.

$$N_2 + 3H_2 \rightleftharpoons 2NH_3$$

Number of molecules on left $= (1 + 3) = 4$
Number of molecule on right $= 2$

So increasing the pressure shifts the equilibrium to the right, increasing the yield of ammonia.

Choosing conditions

For any industrial process, engineers choose the optimum conditions for a reaction by weighing up:

- the effect of temperature on the yield of the product
- the effect of pressure on the yield of the product
- the effects of different factors on reaction rate.

Key words

closed system, dynamic equilibrium

Exam tip

AQA

Remember, lowering temperature increases the yield for the exothermic process and decreases the yield for the endothermic process.

Questions

1 Explain what an equilibrium reaction is. Give an example and an equation to illustrate your answer.

2 Explain the effect of changing the pressure on an equilibrium reaction in the gas phase.

Working to Grade E

1 Complete the table to show the sources of the raw materials for the Haber process.

Raw material	Source
nitrogen	
hydrogen	

2 Write the conditions that are usually chosen for the Haber process in the table below.

temperature (°C)	
pressure (atmospheres)	
catalyst	

Working to Grade C

3 The diagram shows how ammonia is made in the Haber process. Write the letters of the labels below in the correct boxes. You may write one or more letters in each box.

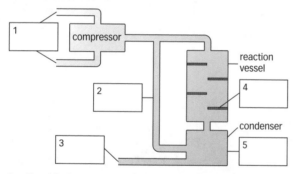

A Purified raw materials enter the apparatus here.

B Unused nitrogen and hydrogen travel through these pipes so that they can be added to the reaction mixture again.

C Liquid ammonia is removed here.

D The reaction mixture is cooled here.

E Iron catalyst.

4 The equation below summarises the processes that occur in the Haber process.
$$N_2(g) + 3H_2(g) \rightleftharpoons 2NH_3(g)$$
Write **T** next to the statements below that are true. Write corrected versions of the statements that are false.

a The symbol $\rightleftharpoons$ shows that the reaction is reversible.

b In the reaction vessel, ammonia molecules break down to make nitrogen and steam.

c In the reaction vessel, nitrogen and hydrogen react together to make ammonia.

d One mole of nitrogen reacts with three moles of hydrogen to make six moles of ammonia.

e There are three gases in the reaction vessel.

Working to Grade A*

5 The equation below represents the equilibrium between two gases – nitrogen dioxide and dinitrogen tetroxide.
$$N_2O_4(g) \rightleftharpoons 2NO_2(g) \; \Delta H = +58 \; KJ/mol$$
Tick the boxes in the table to show how changing the temperature and pressure affect the position of the equilibrium.

Change	Effect on position of equilibrium		
	Shifts left	No change	Shifts right
increasing temperature			
decreasing pressure			
adding a catalyst			

6 For each change in the table above, explain **why** the position of the equilibrium changes (or does not change) in the way you have indicated.

7 Write **T** next to each statement that is true for a system at equilibrium. Write corrected versions of the statements that are false.

a At equilibrium, the rate of the forward reaction is the same as the rate of the backward reaction.

b Each reactant and product is present in the equilibrium mixture.

c At equilibrium, the amounts of products in the mixture gradually increase.

d At equilibrium, the amounts of the reactants in the mixture gradually decrease.

e Equilibrium can be approached from the product side or the reactant side.

f Equilibrium can only be reached in a closed container.

8 The graph shows the relationship between pressure and yield for the Haber process at 450 °C.

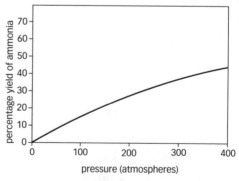

a Describe the relationship shown by the graph.

b Explain why a pressure of 200 atm is chosen for the Haber process.

Examination questions
The production of ammonia

1 Ammonia is a vital chemical. It is used to make fertilisers.

The chemical industry uses the Haber process to manufacture huge amounts of ammonia.

a The Haber process is based on the reaction below.

$$\text{nitrogen} + \text{hydrogen} \rightleftharpoons \text{ammonia}$$

   i Explain what the symbol $\rightleftharpoons$ tells you about the reaction.

....
(1 mark)

   ii The symbol equation for the reaction is given below. It is not balanced.

Write one number on each dotted line to balance the equation.

$$N_2\,(g) + \text{.........}H_2(g) \rightleftharpoons \text{.........}NH_3\,(g)$$

(1 mark)

b The diagram below summarises the stages of the Haber process.

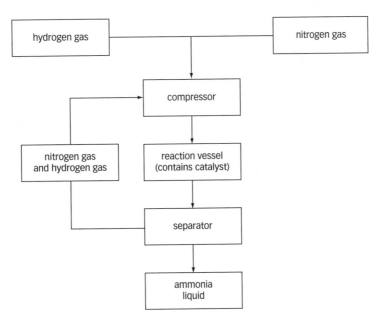

   i Explain why unreacted nitrogen and hydrogen are recycled.

....

....
(2 marks)

   ii Give the purpose of the catalyst.

....
(1 mark)

iii In the Haber process, energy is transferred from the cooling gases in the condenser to the reaction vessel.

Give **two** reasons for this.

1 ....

2 ....

(2 marks)

c The graph shows the relationship between temperature, pressure, and yield of ammonia.

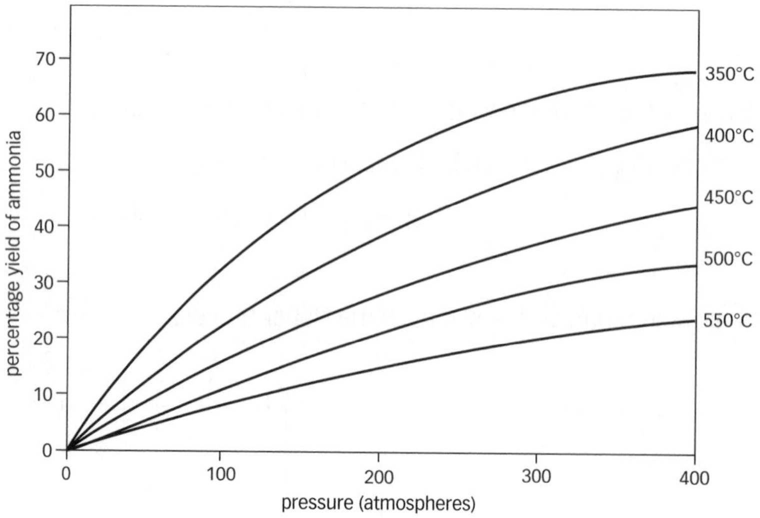

i Use the graph to describe the relationship between **temperature** and **yield** of ammonia.

....

(1 mark)

ii Use your answer to part **ci**, and your own knowledge, to explain why a temperature of 450 °C is often chosen for the Haber process.

In your answer, refer to economic factors and energy use.

....

....

....

(3 marks)

(Total marks: 11)

2 Every year, chemical plants all over the world produce a total of around 150 million tonnes of sulfuric acid.

The equilibrium reaction below is an important stage in the process of the manufacture of sulfuric acid.

sulfur dioxide + oxygen ⇌ sulfur trioxide

a Which of the following statements about the equilibrium reaction are true?

Tick boxes to show the best answers.

In the equilibrium mixture, sulfur dioxide and oxygen are reacting to make sulfur trioxide. ☐

The amount of sulfur dioxide in the equilibrium mixture decreases all the time. ☐

The forward reaction is faster than the backward reaction. ☐

The amount of sulfur trioxide in the equilibrium mixture increases all the time. ☐

In the equilibrium mixture, sulfur trioxide is decomposing to make sulfur dioxide and oxygen. ☐

The backward reaction is faster than the forward reaction. ☐

(2 marks)

b Look at the balanced equation below.

$$2SO_2(g) + O_2(g) \rightleftharpoons 2SO_3 \; \Delta H = -197 \, KJ/mol$$

Use the equation and the energy change for the reaction to predict and explain the effects on the yield of sulfur trioxide of changing the temperature and pressure.

In this question you will get marks for using good English, organising information clearly, and using scientific words correctly.

....

....

....

....

....

....

....

....

....

(6 marks)

(Total marks: 8)

Revision objectives

- ✓ recognise alcohols from their names or formulae
- ✓ describe the reactions of alcohols
- ✓ evaluate the social and economic advantages and disadvantages of the uses of alcohols

Student book references

3.28 Alcohols – 1

3.29 Alcohols – 2

Specification key

✓ C3.6.1

What's in an alcohol?

Alcohols are **organic compounds**. This means that their molecules are made up mainly of carbon and hydrogen atoms.

There are many **alcohols**. They all include the reactive –OH group. A reactive group of atoms in an organic molecule is called a **functional group**.

The alcohols in the table below are members of the same **homologous series**. The members of a homologous series have the same functional group, but different numbers of carbon atoms.

Name	Molecular formula	Structural formula
methanol	CH_3OH	
ethanol	CH_3CH_2OH	
propanol	$CH_3CH_2CH_2OH$	

Properties of alcohols

Methanol, ethanol, and propanol are in the same homologous series. They have similar properties.

- They dissolve in water to form neutral solutions (pH 7).
- They react with sodium. The products are a salt and hydrogen. For example:

  methanol + sodium → sodium methoxide + hydrogen

  ethanol + sodium → sodium ethoxide + hydrogen

- They burn in air. The products are carbon dioxide and water. For example:

$$\text{ethanol} + \text{oxygen} \rightarrow \text{carbon dioxide} + \text{water}$$
$$CH_3CH_2OH + 3O_2 \rightarrow 2CO_2 + 3H_2O$$
$$\text{propanol} + \text{oxygen} \rightarrow \text{carbon dioxide} + \text{water}$$
$$2CH_3CH_2CH_2OH + 9O_2 \rightarrow 6CO_2 + 8H_2O$$

Using alcohols

Alcoholic drinks

Ethanol is the alcohol in alcoholic drinks. Alcoholic drinks have social and economic advantages and disadvantages.

	Advantages	Disadvantages
Social	• Makes people feel relaxed for a short time.	• Slow reaction time – increases risk of road accidents. • Make people forgetful, confused, and more likely to act foolishly. • Cause vomiting, unconsciousness, death.
Economic	• Profitable for drinks companies. • Taxes provide income for government.	• Treating alcohol-related health problems is costly. • Dealing with alcohol-related crime is costly.

Alcohols as solvents

Ethanol and methanol are useful solvents. Ethanol is used as a solvent in perfumes and deodorants, medicines, and food flavourings.

Alcohols as fuels

When alcohols burn in air, much energy is released as heat. So alcohols are useful fuels. There are advantages and disadvantages of using ethanol as a fuel instead of fossil fuels, such as petrol and diesel.

	Advantages	Disadvantages
Environmental	• The crops from which ethanol is manufactured take in carbon dioxide gas as they grow. Some people say this means that ethanol fuel is **carbon neutral**.	• On burning, alcohols produce carbon dioxide gas. • Making fertilisers for the crops produces carbon dioxide. Some people say this means that ethanol fuels are not carbon neutral.
Social	• Made from renewable resources such as sugar cane or maize.	• Crops from which ethanol is made are grown on land that could be used to grow food.

Oxidation reaction

Ethanol can be oxidised to ethanoic acid. It is oxidised by:

• chemical oxidising agents, for example, potassium dichromate(VI) solution
• the action of microbes.

Ethanoic acid is the main acid in vinegar.

Key words

organic compound, alcohol, functional group, homologous series, carbon neutral

Exam tip

You need to be able to write balanced chemical equations for the *combustion* reactions of alcohols, but not for any of their other reactions.

Questions

1 Write the functional group that is in all alcohols.

2 Draw a table to summarise the chemical reactions of ethanol.

3 Write a balanced symbol equation for the combustion reaction of methanol.

What's in a carboxylic acid?

Methanoic acid, ethanoic acid, and propanoic acid are members of the same homologous series. They are all **carboxylic acids**.

Every carboxylic acid molecule includes the functional group –COOH. The atoms in the functional group are arranged like this:

The table shows the formulae of three carboxylic acids.

Name	Molecular formula	Structural formula
methanoic acid	HCOOH	
ethanoic acid	CH_3COOH	
propanoic acid	CH_3CH_2COOH	

Properties of carboxylic acids

Methanoic acid, ethanoic acid, and propanoic acid have similar properties.

- They dissolve in water to form acidic solutions (with pH less than 7).
- They react with carbonates. The products are a salt, carbon dioxide, and water. For example:

  ethanoic acid + sodium carbonate → sodium ethanoate + carbon dioxide + water

- They react with alcohols to make esters.

Using carboxylic acids

Ethanoic acid is the main acid in vinegar. Citric acid is added to some foods and drinks to give them a sour taste.

Many fruits and vegetables contain ascorbic acid, vitamin C. This is vital for health. Vitamin tablet companies include ascorbic acid in their tablets.

The medicine aspirin is also a carboxylic acid. It is a painkiller. Aspirin also reduces blood clotting, so is taken by some people at risk of heart attacks.

H Weak acids

When ethanoic acid dissolves in water, some of its molecules split up. Two ions are formed:

- a positive hydrogen ion
- a negative ethanoate ion.

The ethanoic acid molecule has **ionised**.

Ethanoic acid is a **weak acid** because fewer than 1% of its molecules ionise when they dissolve in water. The equilibrium for the solution of ethanoic acid lies to the left:

$$CH_3COOH(l) + (aq) \rightleftharpoons CH_3COO^-(aq) + H^+(aq)$$

All carboxylic acids are weak acids.

Hydrochloric acid also ionises when it dissolves. All its molecules split up to form hydrogen ions and chloride ions:

$$HCl(g) + (aq) \rightarrow H^+(aq) + Cl^-(aq)$$

Hydrochloric acid is a **strong acid**, because all its molecules ionise when it dissolves.

Weak acids and pH

pH values measure the acidity of a solution. The lower the pH, the more acidic the solution. The pH of a solution tells us about the concentration of hydrogen ions – the greater the concentration of hydrogen ions, the lower the pH.

The table gives the pH of two acids of the same concentration.

hydrochloric acid, 0.1 mol/dm³	pH 1.0
ethanoic acid, 0.1 mol/dm³	pH 2.9

The ethanoic acid has a smaller concentration of hydrogen ions in solution. This is because only 1% of the ethanoic acid molecules are ionised. The ethanoic acid has a higher pH.

Questions

1 Draw a table to summarise the reactions of the carboxylic acids.

2 List **four** uses of carboxylic acids.

3 H Predict and explain the difference in pH of a 0.5 mol/dm³ solution of hydrochloric acid and a solution of ethanoic acid of the same concentration.

Revision objectives

- recognise esters from their names or structural formulae
- describe how esters are made
- identify some physical properties of esters
- recall two uses of esters

Student book references

3.32 Esters

Specification key

✔ C3.6.3

Key words

esters, volatile, esterification

Exam tip

AQA

You need to be able to name one ester – ethyl ethanoate. You should also be able to recognise a compound as an ester from its name or structural formula.

Questions

1 Draw the structural formula of ethyl ethanoate.

2 Write a word equation for the reaction of ethanol with ethanoic acid to form an ester.

3 Write down **two** uses of esters.

What's in an ester?

Esters include the functional group –COO– in their molecules. The atoms in the functional group are arranged like this:

$$-\underset{\underset{O}{\|}}{C}-O-$$

There are many esters. One of them is ethyl ethanoate. Its structural formula is below.

ethyl ethanoate

Properties of esters

Esters share many properties. They:

- are **volatile,** meaning that they easily form vapours
- have distinctive smells – for example, the ester pentyl ethanoate contributes to the smell of pears.

Using esters

The smells and tastes of esters make them useful for:

- making perfumes, shampoos, and shower gels
- flavouring foods such as sweets and chocolates.

Making esters

You can make esters in the laboratory by reacting a carboxylic acid with an alcohol. The process is called **esterification**. For example, ethanoic acid reacts with ethanol to make ethyl ethanoate. You need to use an acid catalyst – concentrated sulfuric acid works well.

The diagrams show what to do.

① 2 cm³ ethanol
1 cm³ concentrated ethanoic acid
three drops of concentrated sulfuric acid

② test tube holder
heat gently

③ pour onto water and smell carefully
ester
watch glass
water

The equation below summarises the reaction in which ethyl ethanoate is made from ethanol and ethanoic acid.

ethanoic acid   ethanol   ethyl ethanoate   water

1 Draw lines to match each compound to one use.

Compound
ethanol
ethanoic acid
pentyl pentanoate

Use
alcoholic drinks
flavouring
vinegar

2 Highlight the one correct word in each pair of **bold** words.

Alcohols dissolve in water to form **neutral/acidic** solutions. Alcohols react with sodium to produce **hydrogen/water**. Carboxylic acids dissolve in water to form **neutral/acidic** solutions. Carboxylic acids react with carbonates to produce **hydrogen/carbon dioxide**.

3 Complete the table below.

Name of group of organic compounds	Examples
alcohols	• ethanol •
carboxylic acids	• •
	• ethyl ethanoate •

4 Use the words in the box below to complete the sentences that follow. Each word may be used once, more than once, or not at all.

> homologous propanoic microbes ethanoic
> solvents fuels perfumes functional propanol
> ethanol oxidising vinegar flavourings drinks

Alcohol molecules contain the _____ group –OH. Methanol, ethanol, and _____ are members of the same _____ series. Alcohols are used as _____, _____ and in _____. The main alcohol in alcoholic drinks is _____. Ethanol can be oxidised to _____ acid either by chemical _____ agents or by the action of _____.

5 Write **T** next to the sentences below that are true. Write corrected versions of the sentences that are false.
  a Carboxylic acids have the functional group –COOOH
  b Carboxylic acids react with acids to produce esters.
  c The molecular formula of methanoic acid is CH_3COOH.
  d Ethanoic acid is the main acid in vinegar.
  e Vinegar is acidic.
  f Propyl propanoate is a carboxylic acid.

6 Fill in the empty boxes.

Name	Molecular formula	Structural formula
methanol		
	CH_3CH_2OH	
	HCOOH	
propanoic acid		

7 For each list below, circle **two** compounds that react together to make an ester.
  a Ethanol, propane, propanoic acid, ethyl propanoate.
  b Propane, propanol, ethanoic acid, ethane.
  c Water, methanoic acid, methane, methanol.
  d Ethanoic acid, ethyl ethanoate, ethanol, ethane.

8 Complete the word equations below.
  a ethanol + oxygen → _____ + water
  b methanol + sodium →
     sodium methoxide + _____
  c ethanol + sodium → _____ + _____
  d propanol + oxygen → _____ + _____
  e ethanoic acid + sodium carbonate →
     sodium ethanoate + _____ + water
  f propanoic acid + calcium carbonate →
     _____ + carbon dioxide + _____
  g _____ + ethanoic acid →
     ethyl ethanoate + water
  h propanol + propanoic acid → _____ + water

9 Balance the equations below.
  a $CH_3OH + O_2 \rightarrow CO_2 + H_2O$
  b $CH_3CH_2OH + O_2 \rightarrow CO_2 + H_2O$

10 Explain the meaning of the statement below:
    Ethanoic acid is a weak acid.

11 Matthew measures the pH of three acids, A, B, and C. Each acid has the same concentration. His results are in the table.
Matthew knows that one of the acids is hydrochloric acid, and that the other two are carboxylic acids.

Acid	pH
A	1.2
B	3.7
C	4.1

  a Which acid is hydrochloric acid? Explain how you decided.
  b Is acid C a stronger or weaker acid than acid B? Explain how you decided.

1 The solvent in this nail varnish is ethyl ethanoate.

a Which formula represents the structure of ethyl ethanoate?

Draw a ring around the correct formula.

(1 mark)

b To make ethyl ethanoate, scientists react two substances together.

Tick the names of the **two** chemical families to which these substances belong.

alcohols

esters

carboxylic acids

alkanes

(1 mark)

c Sulfuric acid is added to the reaction mixture during the synthesis of ethyl ethanoate. Explain why.

....

(1 mark)

d Pentyl pentanoate has the same functional group as ethyl ethanoate.

It is used as a food flavouring.

Write down the property of pentyl pentanoate that makes it suitable for use as a food flavouring.

....

(1 mark)
(Total marks: 4)

2 Identify **three** uses of alcohols.

Choose **one** of these uses, and evaluate the social and economic advantages and disadvantages of this use.

In this question you will get marks for using good English, organising information clearly, and using scientific words correctly.

....

....

....

....

....

....

....

....

....

....

(6 marks)
(Total marks: 6)

3 This question is about hexanoic acid.

a Complete the diagram below to show the structural formula of hexanoic acid.

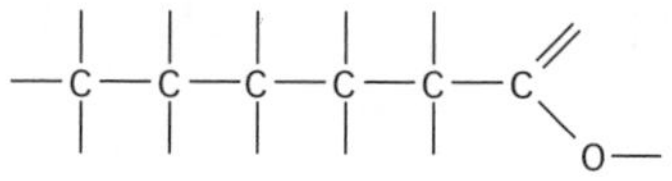

(1 mark)

b Hexanoic acid reacts with calcium carbonate.

Complete the word equation for the reaction below.

hexanoic acid + calcium carbonate → calcium hexanoate + ... + water

(1 mark)

c A student tests the pH of four organic compounds.

Her results are in the table.

Compound	pH
A	4.3
B	7.0
C	10.0
D	12.2

Which of the compounds in the table could be hexanoic acid?

....

(1 mark)

H

d i Explain what makes hexanoic acid a weak acid.

....

....

(2 marks)

ii A student has samples of the acids listed below.

Predict which of the acids has the highest pH value.

Explain your choice.

List of acids
0.1 mol/dm^3 hydrochloric acid
0.1 mol/dm^3 sulfuric acid
0.1 mol/dm^3 nitric acid
0.1 mol/dm^3 ethanoic acid

....

....

(1 mark)

(Total marks: 6)

Designing an investigation and making measurements

In this module there are several opportunities to design investigations and make measurements. These include investigating the volumes of acids and alkalis that react together, and measuring the pH of solutions of alcohols and carboxylic acids.

As well as demonstrating your investigative skills practically, you are likely to be asked to comment on investigations done by others. The examples below offer guidance in these skill areas. They also give you the chance to practise using your skills to answer the sorts of question that may come up in exams.

The volumes of acid and alkali solutions that react with each other

Suzanna tested the hypothesis that the higher the concentration of hydrochloric acid, the smaller the volume required to neutralise an alkali. She did titrations to measure the volumes of hydrochloric acid of different concentrations that reacted with sodium hydroxide solution. She used the apparatus below.

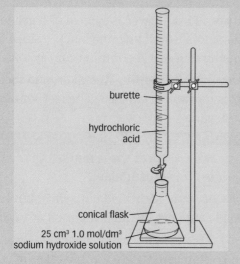

burette

hydrochloric acid

conical flask

25 cm³ 1.0 mol/dm³ sodium hydroxide solution

1   Suzanna did a trial run before she started her main investigation. Suggest **three** questions Suzanna might have wanted to consider in her trial run.

In general, a trial run helps an investigator to select appropriate values to be recorded.

In your answer to this question, suggest specific questions Suzanna might have wanted to address before starting her main investigation, such as 'what volume of acid shall I use?'.

2   Suzanna's results are in the table.

Concentration of hydrochloric acid (mol/dm³)	Volume of acid required to neutralise 25 cm³ of 1.0 mol/dm³ sodium hydroxide solution (cm³)			
	Run 1	Run 2	Run 3	Mean
0.5	49.50	49.60	49.40	49.50
0.8	31.05	31.00	30.95	31.00
1.0	25.15	25.10	25.05	25.10
1.5	17.20	16.60	16.00	16.60
2.0	12.50	12.50	12.50	12.50

Identify the independent and dependent variables.

- The independent variable is the one that is changed or selected by the investigator.
- The dependent variable is measured for each change in the independent variable.

3   Describe how Suzanna can make sure her measurements are valid.

For a measurement to be valid it must measure only the appropriate variable – how can Suzanna ensure she does this? One answer is to keep the volume of acid the same in each run. Can you think of others?

4   Describe how Suzanna can make sure she does a fair test.

In a fair test, only the independent variable should affect the dependent variable. All the other variables must be kept the same.

In your answer, include a list of the control variables.

5   Compare the precision of the data obtained when the concentration of hydrochloric acid was 1.0 mol/dm³ and when the concentration of hydrochloric acid was 1.5 mol/dm³.

Data is precise if values obtained by repeated measurements cluster closely. In Suzanna's investigation:
- If the values obtained for a certain acid concentration are close together, the measurements for this concentration are precise.
- If the values for a certain acid concentration are not close together, the measurements for this concentration are not precise.

6   Suggest why the values obtained for repeated readings are not exactly the same.

There will always be some variation in the actual value of a variable, no matter how hard an investigator tries to repeat an event. In answering this question, think about what Suzanna might have done to have obtained results that are slightly different, or how the equipment she used might have led to slightly different results for any given acid concentration.

AQA *Upgrade*

Answering a question with data response

In this question you will be assessed on using good English, organising information clearly, and using specialist terms where appropriate.

1 A student wants to identify two salts. The table below shows the tests he did, and gives his results. Identify salts A and B, using data from the table to explain and support your decisions. *(6 marks)*

Test number	Test	Salt A observations	Salt B observations
1	Add sodium hydroxide solution.	blue precipitate	green precipitate
2	Add dilute hydrochloric acid to a sample of the solid salt.	fizzing – gas produced that made limewater cloudy	no change
3	Dissolve a little of the salt in water. Add barium chloride solution and hydrochloric acid.	no change	precipitate formed on filtering, could see that precipitate was white
4	Dissolve a little of the salt in water. Add silver nitrate solution and nitric acid.	no change	no change

QUESTION

G–E

B is iron (III) chloride – this is wot the test reasults show me. And A contains copper. And B has chloride ions in it. But I dont no wot the other bit of A is. I think he needs to do some more tests if he has time he could try tests with pottasium and sodium these are fun and can eksplode.

Examiner: This answer is typical of a grade-G candidate. It is worth just one mark, gained for recognising that salt A contains copper ions.

There are several spelling mistakes, and the answer is not well organised. The last sentence is not relevant.

D–C

A is copper (II) carbonate. I know it has carbonate because the solid fizzed when he drip acid onto it and the gas made limewater milky because it is carbon dioksyd.

B is definitely iron (II) sulfate. I be sure about this becaus of the test reasults.

Examiner: This answer is worth three marks out of six. It is typical of a grade-C or -D candidate.

The candidate has correctly identified both salts, but has only given reasons to support the identification of one ion (carbonate).

The answer is well organised, with a few mistakes of grammar and spelling.

B–A*

Salt A is copper (II) carbonate. The test with sodium hydroxide shows that the salt contains copper (II) ions. Test number two shows that salt A is a carbonate, since when it reacted with acid it made a gas that made limewater cloudy (carbon dioxide). Test 1 shows that salt B contains iron (II) ions. The green precipitate is iron (II) hydroxide. Test 3 shows that salt B also contains sulfate ions. The precipitate is barium sulfate.

The student had to filter the mixture for the barium chloride test, otherwise he would not have known that the precipitate was white, since the solution of the salt in water must have been coloured. Together, tests 1 and 3 show that salt B is iron (II) sulfate.

Examiner: This is a high-quality answer, typical of an A* candidate. It is worth six marks out of six.

The candidate has correctly identified the two salts, and given clear and detailed reasons to support his decisions.

The answer is well-organised. The spelling, punctuation, and grammar are faultless. The candidate has used several specialist terms.

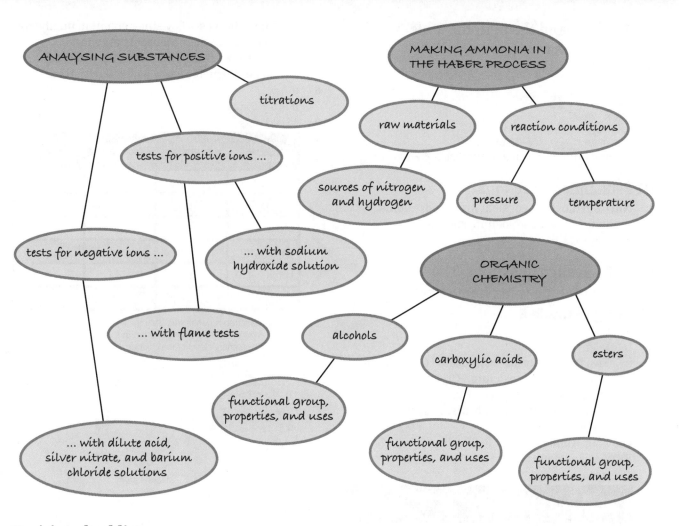

ANALYSING SUBSTANCES

titrations

tests for positive ions ...

tests for negative ions ...

... with sodium hydroxide solution

... with flame tests

... with dilute acid, silver nitrate, and barium chloride solutions

MAKING AMMONIA IN THE HABER PROCESS

raw materials

reaction conditions

sources of nitrogen and hydrogen

pressure

temperature

ORGANIC CHEMISTRY

alcohols

carboxylic acids

esters

functional group, properties, and uses

functional group, properties, and uses

functional group, properties, and uses

Revision checklist

- Many metal compounds give distinctive flame colours, including compounds of lithium (crimson), sodium (yellow), potassium (lilac), calcium (red), and barium (green).
- You can identify some metal ions in solution by adding sodium hydroxide solution. The precipitates formed have distinctive colours.
- You can test for carbonates by adding dilute acid, for halide ions by adding silver nitrate solution, and for sulfates by adding barium chloride solution.
- You can use titrations to measure the volumes of acid and alkaline solutions that react with each other.
- If you know the concentration of one reactant, you can use titration results to calculate the concentration of the other reactant.
- The raw materials for the Haber process are nitrogen (obtained from the air) and hydrogen (often obtained from natural gas).
- In the Haber process, nitrogen and hydrogen react together to form ammonia, NH_3, in a reversible reaction. The reaction takes place at high temperature (450 °C), high pressure (200 °C), and in the presence of an iron catalyst.

- When a reversible reaction occurs in a closed system, equilibrium is reached when the reactions occur at the same rate in each direction.
- The relative amounts of the reacting substances at equilibrium depend on the conditions of the reaction. Changing the temperature or pressure changes the amounts of substances at equilibrium.
- Alcohols have the functional group –OH. They form neutral solutions in water, react with sodium to produce hydrogen, and burn in air. Ethanol is oxidised to ethanoic acid.
- Alcohols are used as fuels, solvents, and in drinks.
- Carboxylic acids have the functional group –COOH. They form acidic solutions in water, react with carbonates to produce carbon dioxide, and react with alcohols to produce esters.
- Ethanoic acid is the main acid in vinegar.
- Carboxylic acids are weak acids because they do not ionise completely in water.
- Esters have the functional group –COO–. They are volatile compounds with distinctive smells.
- Esters are used as flavourings and perfumes.

C1 1: Atoms and the periodic table

1 Proton +1, neutron 0, electron –1.

2

3 Atomic number = 9, mass number = 9 + 10 = 19.

C1 2: Chemical reactions

1 Nitrogen dioxide – covalent; iron oxide – ionic.
2 sodium + chlorine → sodium chloride
3 $2Li + F_2 \rightarrow 2LiF$

C1 1–2 Levelled questions: The fundamental ideas in chemistry

Working to Grade E

1 a O
  b Iron
  c Five from: chlorine – Cl; cobalt – Co; cadmium – Cd; carbon – C; calcium – Ca; caesium – Cs; chromium – Cr; copper – Cu.
  d Four from: helium, neon, argon, krypton, xenon, radon.
  e Four from: Li, Na, K, Rb, Cs, Fr.
2 Protons and neutrons.
3 100, groups.
4

Name of element	Number of electrons	Electronic structure
lithium	3	2.1
sodium	11	2.8.1
potassium	19	2.8.8.1

5 One from: they react vigorously with water to form a hydroxide and hydrogen gas; they react vigorously with oxygen from the air to form oxides.

6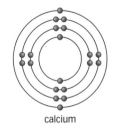

carbon   silicon

calcium

Working to Grade C

7 a 8
  b Helium

c The noble gases have the maximum number of electrons in their highest energy level/outermost shell. This electronic structure makes the noble gases particularly stable.

8 a 5
  b 5 + 6 = 11
9 45 – 21 = 24
10

Name of element	Number of protons	Number of neutrons	Number of electrons
hydrogen	1	0	1
oxygen	8	8	8
sodium	11	12	11
aluminium	13	14	13
boron	5	6	5
calcium	20	20	20

11 one electron is transferred from the highest energy level of the sodium atom to the highest energy level of the chlorine atom

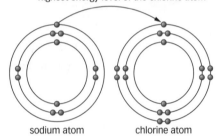

sodium atom   chlorine atom

12 Covalent bonding.
13 Ions
14 a carbon + oxygen → carbon dioxide
  b magnesium + chlorine → magnesium chloride
  c iron + sulfur → iron sulfide
  d sulfuric acid + sodium hydroxide → sodium sulfate + water
  e lead carbonate → lead oxide + carbon dioxide
15 6.4 g – 3.2 g = 3.2 g

Working to Grade A*

16 a Oxygen: $2Mg + O_2 \rightarrow 2MgO$
  b Balanced
  c Sodium and hydrogen: $H_2SO_4 + 2NaOH \rightarrow Na_2SO_4 + 2H_2O$
  d Balanced
  e Balanced
  f Oxygen and hydrogen: $CH_4 + 2O_2 \rightarrow CO_2 + 2H_2O$

C1 1–2 Examination questions: The fundamental ideas in chemistry

1 a i The atom of the element has 6 electrons in its highest energy level. (1)
    ii 8 (1)
  b i 9 (1)
    ii 9 + 10 = 19 (1)
    iii Fluorine (1); F (1).

iv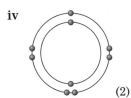

(2)

2 **a** **i** lithium + oxygen → lithium oxide (1)
ii $4Li + O_2 \rightarrow 2Li_2O$ (1)
b 60 g (2)
c **i** metal (1); loses (1); positive (1)
ii 2 (1)
d Helium (1)
e Number of protons = 3 (1);
Number of neutrons = 4 (1)

C1 3: Limestone and building materials
1 Mortar – cement and sand; concrete – cement, sand, aggregate.
2 A suspension of calcium carbonate in water.
3 **a** calcium oxide + water → calcium hydroxide
b magnesium carbonate + nitric acid → magnesium nitrate + carbon dioxide + water

C1 3 Levelled questions: Limestone and building materials
Working to Grade E
1 **a** One from: limestone makes buildings and cement, quarries provide jobs.
b Quarries create extra traffic.
c Old quarries can be made into lakes.
d Tourists may stop visiting an area with a new quarry.
2 clay, mortar, concrete.
3 potassium carbonate
4 Take a sample of colourless limewater. Bubble the gas through the limewater. If the gas is carbon dioxide, the limewater will start to look cloudy.
5

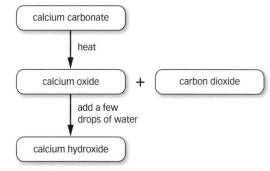

Working to Grade C
6 Calcium carbonate → calcium oxide + carbon dioxide
7 **a** zinc carbonate → zinc oxide + carbon dioxide
b lead carbonate → lead oxide + carbon dioxide
8 **a** Copper sulfate
b Zinc chloride
c Calcium nitrate
d Magnesium chloride
e Sodium sulfate

9 **a** → sodium nitrate + carbon dioxide + water
b → magnesium nitrate + carbon dioxide + water
c → calcium chloride + carbon dioxide + water
d → zinc sulfate + carbon dioxide + water
10 **a** Mild steel
b Oak wood and limestone conduct heat least well. This means they are good insulators of heat, which helps to prevent heat loss from buildings made of these materials.
c There is no data for the limestone, and the high strength concrete cracks if it is pulled.

Working to Grade A*
11 $ZnCO_3 \rightarrow ZnO + CO_2$
12 $Ca(OH)_2 + CO_2 \rightarrow CaCO_3 + H_2O$
13 **a** $MgCO_3 + 2HNO_3 \rightarrow Mg(NO_3)_2 + CO_2 + H_2O$
b $ZnCO_3 + 2HCl \rightarrow ZnCl_2 + CO_2 + H_2O$
c $Na_2CO_3 + HCl \rightarrow 2NaCl + CO_2 + H_2O$
d Balanced.
e $CaCO_3 + 2HNO_3 \rightarrow Ca(NO_3)_2 + CO_2 + H_2O$

C1 3 Examination questions: Limestone and building materials
1 **a** **i** Quarry companies can sell the limestone (1).
ii Quarries make the land unavailable for other purposes (1).
b **i** calcium carbonate → calcium oxide + carbon dioxide (1)
ii Traditional lime kilns were open to the air so that the carbon dioxide produced in the decomposition reaction could escape (1).
iii Calcium hydroxide solution is also known as limewater. It is used to test for carbon dioxide gas. It is also used to neutralise acids, for example in lakes whose pH is too low (1 mark for each correct answer).
c Calcium sulfate (1) and magnesium sulfate (1).
d $CaCO_3 + 2HCl \rightarrow CaCl_2 + CO_2 + H_2O$ (1)
2 **a** **i** Distance of Bunsen flame from bottom of test tube (1); amount of limewater (1).
ii Time for limewater to begin to look cloudy (1).
b Zinc carbonate → zinc oxide + carbon dioxide (1)
c magnesium oxide (1) + carbon dioxide (1)
d calcium hydroxide + carbon dioxide → calcium carbonate + water (1)

C1 4: Extracting metals
1 Mine the ore; concentrate the ore to separate its metal compounds from waste rock; extract the metal from its compounds; purify the metal.
2 Aluminium is more reactive than carbon so it cannot be extracted by reduction with carbon. Instead, an electric current is passed through molten aluminium oxide.

3 Phytomining – planting certain plants on low-grade copper ores. The plants absorb copper compounds. Burn the plants. The ash is rich in copper compounds.

Bioleaching – use bacteria to make solutions of copper compounds. Use electrolysis or displacement reactions to extract copper from these solutions.

C1 5: Metals: recycling, properties, alloys, and uses

1 Reasons for recycling metals – some ores are in short supply; extracting metals from ores creates waste that may damage the environment; extracting metals from ores requires electrical energy or heat energy.

2 Low carbon steel is easily shaped and is used for food cans and car body panels. High carbon steel is hard and is used to make tools. Stainless steel is resistant to corrosion and is used to make cutlery and surgical instruments.

3 Titanium is suitable for aeroplanes because it has a low density and does not corrode. Titanium is suitable for artificial hip bones because it does not corrode.

C1 4–5 Levelled questions: Metals and their uses

Working to Grade E

1 Gold
2 b, c, d, a.
3 Molybdenum, manganese, scandium, titanium.
4 Recycling produces smaller amounts of waste materials, and requires less energy.
5 a

Type of steel	Properties
Stainless steel	Resistant to corrosion
High carbon steel	Very hard
Low carbon steel	Easily shaped

b Stainless steel – cutlery or surgical instruments.
High-carbon steel – tools.
Low-carbon steel – food cans or car body panels.
6 brittle, cast iron, compression.
7 a It has a low density and is resistant to corrosion.
b It resists corrosion, even in salty water.
8 Reduced means that oxygen has been removed from a compound. Reducing iron oxide with carbon produces iron and carbon monoxide and/or carbon dioxide.

Working to Grade C

9 a tin, lead.
b Two from: calcium, magnesium, aluminium.
c Gold
10 It is expensive to extract titanium from its ore because the extraction has many stages and requires much energy.

11 Obtaining copper by phytomining or bioleaching creates less waste and requires less energy than extracting copper from low-grade ores by smelting.
12 a Electrode A.
b The negative electrode, electrode B.
13 copper sulfate + iron → copper + iron sulfate
14 a Aluminium alloy 7075.
b This alloy is harder than the other two materials in the table, and has the greatest tensile strength.
c Pure aluminium has a slightly lower density than aluminium alloy 7075, so aluminium alloy 7075 is heavier than pure aluminium per unit volume.

C1 4–5 Examination questions: Metals and their uses

1 a 2 marks from: copper can easily be bent into different shapes without cracking (1); it is hard enough to make pipes and tanks (1); it does not react with water (1).
b i Nickel brass and naval brass. (1)
ii Nickel brass is an alloy of copper, so it is probably harder than pure copper (1).
c i This method creates large amounts of waste (1); requires a great deal of energy (1).
ii Particular species of plant are planted on low-grade copper ore. The plants take in copper compounds (1). The plants are harvested and burnt. Copper is extracted from their ash (1).
2 a i The scientists found the total mass of indium in samples 2–6. They then divided this value by the number of samples (5). The answer was 100 g. They decided not include the value from sample 1, since it is anomalous (1).
ii 40 tonnes (2).
iii 399 960 tonnes (1).
iv The waste may produce a large and unsightly heap (1).
b Any reasoned answer acceptable, for example: I recommend that the mine should be allowed to reopen. The advantages of opening the mine are that 400 jobs could be created. A compound made from indium is used to make solar cells, for which demand is increasing. I think these advantages outweigh the disadvantages such as the fact that indium is harmful if swallowed or breathed in, and that pollution has been linked to the extraction of indium metal in China, but the mine owners must take precautions to minimise the risks from these hazards. (1 mark for each reasoned point made that supports the recommendation.)

C1 6: Crude oil and hydrocarbons

1 A hydrocarbon is a compound made up of hydrogen and carbon only; a mixture consists of two or more elements or compounds that are not chemically joined together; a fraction is a mixture of hydrocarbons whose molecules have a similar number of carbon atoms; an alkane is a hydrocarbon that is saturated.

2 A mixture consists of two or more elements or compounds that are not chemically joined together. Each of the substances in a mixture has its own properties, and being part of a mixture does not affect these properties. You can separate mixtures by physical means such as filtration and distillation.

3 CH_4, C_2H_6, C_3H_8.

C1 7: Hydrocarbon fuels

1 As molecule size increases, boiling point and viscosity increase and flammability decreases.

2

Substance	Produced as a result of ...	Causes these problems ...
Carbon dioxide	The combustion of hydrocarbon fuels	Global warming
Carbon monoxide	Partial combustion of hydrocarbon fuels	Poisonous – can cause death if breathed in
Sulfur dioxide	Combustion of fuels that contain sulfur	Acid rain
Oxides of nitrogen	Combustion of hydrocarbons in air at high temperatures	Acid rain
Particulates	Partial combustion of hydrocarbons	Global dimming

C1 8: Biofuels and hydrogen fuel

1 Biodiesel and diesel both produce carbon dioxide on burning. The plants from which the biofuels were made removed carbon dioxide from the atmosphere during photosynthesis whilst they were growing. But growing biofuel crops require inputs such as fertilisers. The production of fertilisers results in carbon dioxide gas emissions. Overall, biofuels are likely to result in smaller carbon dioxide emissions than diesel.

2 Benefits – only one product, water vapour, which is non-polluting; can be produced from renewable resources, such as methane from animal waste. Drawbacks – since hydrogen gas is explosive with air, it is difficult and dangerous to store and transport.

C1 6–8 Levelled questions: Crude oil and fuels

Working to Grade E

1 a False – crude oil is a mixture of compounds.
  b False – a mixture consists of two or more elements or compounds that are not chemically combined.
  c False – when substances are mixed together, their chemical properties remain unchanged.
  d True
  e True

2 a hydrocarbons
  b hydrogen
  c alkanes
  d saturated

3 d, a, b, e, c, f.

4 Nitrogen dioxide – formed when hydrocarbons burn in air at high temperatures.
  Carbon monoxide – formed when hydrocarbons burn in a poor supply of air.
  Sulfur dioxide – formed when a sulfur-containing fuel burns.

5 a Global warming.
  b Acid rain.
  c Acid rain.
  d Global dimming.

Working to Grade C

6

Name of alkane	Molecular formula	Structural formula
Methane	CH_4	
Ethane	C_2H_6	
Propane	C_3H_8	
Butane	C_4H_{10}	

7 a increases
  b decreases
  c decreases

8 a carbon dioxide + water.
  b methane + oxygen → carbon dioxide + carbon monoxide + water.

9 a $72 - 20 = 52\,°C$
  b Methane
  c They are oxidised.

10

	Ethanol	Petrol
Use of renewable resources	Pro: Plants from which ethanol obtained can be grown each year. Con: Growing the plants from which ethanol is made requires inputs of fertilisers which are made from non-renewable resources.	Pro: None as petrol is non-renewable. Con: Obtained from non-renewable crude oil.
Fuel storage and use	Pro: Relatively safe and easy to store. Con: Storage tanks and fuel pumps not yet widespread at filling stations.	Pro: Relatively safe and easy to store. Con: Because filling stations can only store enough fuel to last a few days, customers are vulnerable if supplies are threatened.
Combustion products	Pro: Produces water on burning. Con: Produces carbon dioxide on burning.	Pro: Produces water on burning. Con: Produces carbon dioxide on burning.

C1 6–8 Examination questions: Crude oil and fuels

1 a Carbon dioxide – global warming (1);
sulfur dioxide and oxides of nitrogen – acid rain (2);
solid particles – global dimming (1).

b i Methane + oxygen → carbon dioxide + water (2)

   ii Carbon monoxide. (1)

2 a evaporate (1); condense (1); cool (1).

b

	Letter	
The place where a mixture of vapours enters the column.	E	(1)
The hottest part of the column.	D or F	(1)
The place where the fraction containing substances with the highest boiling points leaves the column.	F	(1)
The place where the fraction containing the most flammable substances leaves the column.	A	(1)
The place where methane gas leaves the column.	A	(1)

3 a

(2)

b i As the molecule size increases from methane to hexane, so the boiling point increases (1).

   ii One from: pentane, hexane (1).

4 Marks are awarded for using good English, organising information clearly and using specialist terms where appropriate. 6 marks available.
Points to include:
- Hydrogen can be produced from renewable sources, such as methane from animal waste.
- Petrol and diesel are produced from non-renewable crude oil.
- Burning hydrogen produces just one waste product – water vapour.
- Burning petrol and diesel produces carbon dioxide gas, which causes global warming.
- Burning petrol and diesel produces oxides of nitrogen, which causes acid rain.
- Hydrogen is an explosive gas, so the risks associated with storing and transporting it are greater than the risks associated with storing and transporting liquid petrol and diesel.

How Science Works: Atoms, rocks, metals, and fuels

1 Independent variable – fuel; dependent variable – temperature change.

2 Volume of water, mass of oil burned, distance between top of flame and bottom of water container.

3 The resolution of the thermometer that can detect a temperature change of 0.5 °C is greater, so smaller changes in temperature can be detected.

4 Yes, the results support the hypothesis, because the temperature change of the water on burning 1 g of each fuel increases as the number of carbon atoms in a molecule of the fuel increases. The increasing temperature change of the water is a measure of the energy released on burning a fuel.

5 Any value between 46 and 50 °C.

6 To find out how close together the repeated values are – the closer together they are, the closer they are likely to be to the true value; to spot any anomalous data.

C1 9: Obtaining useful substances from crude oil

1 Cracking the naphtha fraction of crude oil produces alkanes with smaller molecules that can be added to the petrol fraction, for which there is a high demand.

2 Conditions – the hydrocarbons must be vapourised and then passed over a hot catalyst, or mixed with steam and then heated to a high temperature.
Products – alkanes (with smaller molecules than those of the reactants) and alkenes such as ethene.

3 Ethene C_2H_4 and propene C_3H_6.

4 Bubble the alkene through orange bromine water, which will become colourless.

C1 10: Polymers

1. Dental polymers are white, hard and tough. They are poor conductors of heat. The fact that they are poor conductors of heat mean that it is not uncomfortable to eat very hot or cold foods. The fact that they are white means that they match the colour of the teeth. The fact that they are hard and tough means that they will be hard-wearing and not need replacing for many years, if at all.
2. Hydrogels absorb huge volumes of liquids. This means they are suitable for nappies, which must absorb large volumes of urine.
3. Environmental advantage – cornstarch bags are biodegradable, whereas poly(ethene) bags persist in the environment for many years; economic advantage – growers of corn and producers of cornstarch bags have products to sell; social – cornstarch is renewable, so it does not take supplies from future generations. Poly(ethene) is produced from products obtained from crude oil, which is a non-renewable resource.

C1 11: Ethanol

1.

	From ethene by hydration	From glucose by fermentation
Raw materials	Crude oil (ethene) and water (steam)	Plants (sugars)
Conditions	Catalyst 300 °C	Enzymes from yeast catalyse the reaction 37 °C
Products	Ethanol	Ethanol and carbon dioxide

2. Making ethanol from ethene uses up a non-renewable resource – crude oil. This is a disadvantage. Another disadvantage is that the energy costs of producing ethanol from ethene are relatively high, since the reaction takes place at a high temperature. The raw materials for making ethanol by the fermentation of plant sugars are renewable, which is an advantage. The energy costs of this process are relatively low, which is also an advantage.

C1 9–11: Levelled questions: Other useful substances from crude oil

Working to Grade E

1. Poly(ethene), petrol, diesel.
2. Dental fillings – hard and tough and a poor conductor of heat; disposable nappies – absorbs large volumes of liquid; breathable waterproof fabrics – allows water vapour to pass through its tiny pores, but not liquid water; to make mattresses that mould to the body – changes shape in response to warming or pressure.

3. a True
   b A molecule of poly(ethene) is made by joining together thousands of ethene molecules.
   c True
   d The monomer propene makes poly(propene).
4. smaller, vapourise, vapours, catalyst, steam, decomposition.

Working to Grade C

5. C_2H_4, ethene

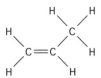

   C_3H_6, propene

6. C_4H_8 and C_7H_{14} are alkenes.
7.

Name of hydrocarbon	Results
Ethane	Bromine water does not change colour
Propene	Colour change from orange to colourless
Ethene	Colour change from orange to colourless
Butane	Bromine water does not change colour

8.

9. The relative demand for the petrol fraction is greater than its typical relative proportion in crude oil. Cracking the naphtha fraction produces hydrocarbons in the petrol fraction, so increasing the amount of petrol an oil company has available to sell.

10.

Starting materials	Type of reaction	Conditions
Glucose	Fermentation	Needs natural catalyst from yeast; 37 °C
Ethene and water	Hydration	Needs catalyst; 300 °C

11

	Advantages	Disadvantages
From ethene	No fertilisers are required, so water pollution is unlikely.	Takes place at a higher temperature, so requires more energy, resulting in more pollution.
From glucose	Takes place at a lower temperature, so requires less energy, resulting in less pollution (for example CO_2 emissions if the heat is supplied by burning a hydrocarbon).	Plants are the raw material for this process. Growing the crops may require inputs of fertilisers, which may cause water pollution.

12 A benefit of plastic recycling is that reserves of non-renewable crude oil are not being used up to produce the plastic. If plastics are recycled, then less plastics are taken to landfill, or incinerated. Incineration produces polluting gases. A problem of plastic recycling is collecting the plastics, which is time-consuming and requires energy inputs for the transporting plastics to a central collection point. A second problem is the sorting of the plastics into different types, which is labour-intensive and so expensive.

C1 9–11 Examination questions: Other useful substances from crude oil

1 **a** Thermal decomposition (1).

b The aluminium oxide acts as a catalyst (1).

c **i** C_2H_4 (1)

   ii Bubble the ethene through orange bromine water (1). The orange bromine water will become colourless (1). This shows that the ethene has a double bond (1).

2 **a** The bags are not biodegradable, which means they are not broken down by microbes (1).

b Jute is made from a plant material, which is renewable, unlike the raw material for making poly(ethene), which is non-renewable crude oil (1).

   Jute bags are biodegradable, so can be broken down by microbes when they are disposed of, unlike poly(ethene) bags, which are non-renewable (1).

3 **a**

(3)

b **i** Two from: poly(butene) does not react with most things that may be mixed with the water, for example detergents and oils. It is flexible, so is easy to put into place under roads and so on. It is elastic, so stretches if the water in the pipes freezes and expands. (1 mark for each)

ii The material of the pipe may react with the chlorine or its compounds that are dissolved in the water (1). The reaction may lead to the pipe being damaged (1).

C1 12: Plant oils and their uses

1 Plant oils can be extracted from seeds, fruits and nuts.

2 The olives are crushed and pressed. The oil is then separated from water and other impurities.

3 Vegetable oils can be used as vehicle fuels since they release large amounts of energy on burning.

C1 13: Emulsions and saturated and unsaturated fats

1 An emulsion is a mixture of oil and water than does not separate out.

2 Add bromine water to the oil in a test tube. Place a bung in the test tube. Shake. If the oil is unsaturated, the mixture will change colour from orange to colourless.

3 Unsaturated vegetable oils can be hardened by adding hydrogen gas to them. The reaction happens at a temperature of about 60 °C. A nickel catalyst is required.

C1 12–13 Levelled questions: Plant oils and their uses

Working to Grade E

1 seeds, nuts, fruit.

2 b, d, a, c

3 **a** Oils do not dissolve well in water.

b True

c Emulsifiers make emulsions more stable.

d A student shakes a mixture of oil, vinegar, and sugar. Afterwards, he sees two layers of liquid. This shows that the sugar is not an emulsifier.

4 **a** D

b A

c E

d B

Working to Grade C

5 **a** Coconut oil and palm oil.

b Coconut oil and palm oil.

6 **a** Catherine, because more energy is released on eating the potatoes cooked by this method. The extra energy comes from the oil absorbed by the potatoes on cooking.

b The advantages of cooking potatoes in oil compared to cooking them in water are that the potatoes cook more quickly in oil, and have a flavour that some people prefer.

c The disadvantage of cooking potatoes in oil is that the resulting food releases more energy on digestion than potatoes cooked in water. Eating lots of oil-rich foods can lead to weight gain.

7 a Eating too much fat can lead to weight gain. Cutting down on fats eaten can help to stabilise or reduce weight.
  b Saturated fats can raise blood cholesterol, which increases the risk of heart disease. Unsaturated fats are better for health.
8 a Independent variable – type of plant oil; dependent variable – temperature increase of water.
  b Control variables – distance of flame from water, volume of water.
  c The oil which causes the greatest water temperature increase per gram of oil is the oil which releases the most energy on burning.

Working to Grade A*
9 End A is hydrophilic, since it is in the water.
10 a Hydrogen
   b Nickel catalyst, 60 °C.
   c The hardened oil boils at a higher temperature than the unsaturated oil from which it is made.
   d Hardened oils are used for spreads and to make cakes and pastries. The fact that they are solid at room temperature makes them suitable for these purposes.

C1 12–13 Examination questions: Plant oils and their uses
1 a Detergent, mustard powder, egg yolk. (1)
  b i It is a viscous liquid with a good coating ability (1).
    ii Two from: paints, cosmetics, ice creams. (2)
  c End A is hydrophobic (1), since it is in the oil (1).
2 Advantages of cooking in oil – cook more quickly (1), taste that many people prefer (1); disadvantage – eating too much oil can cause weight gain (1).
3 a i C and D (1).
    ii zero (1)
  b i Hydrogen is added to the unsaturated vegetable oil (1). A nickel catalyst is required (1), and a temperature of 60 °C (1).
    ii Spreads for bread and as an ingredient for cakes and pastries (1). They are suitable for these purposes because they are solids at room temperature (1).

C1 14: Changes in the Earth
1 The Earth's crust, the atmosphere, and the oceans.
2 Tectonic plates move at speeds of a few centimetres a year.
3 The Earth's mantle moves because, deep inside the Earth, natural radioactive processes heat the mantle. The heat drives convection currents within the mantle.
4 The shapes of Africa and South America look as if they might once have fitted together; there are fossils of the same plants on both continents. There are rocks of the same type at the edges of the two continents where they might once have been joined.

C1 15: The Earth's atmosphere 1
1 Nitrogen – about 80%; oxygen – about 20%, smaller proportions of carbon dioxide, water vapour, the noble gases.
2 Plants removed carbon dioxide from the atmosphere in photosynthesis; carbon dioxide gas dissolved in the oceans.
3 No one was around to make observations at the time.

C1 16: The Earth's atmosphere 2
1 There is more carbon dioxide in the atmosphere as a result of humans burning fossil fuels. This leads to more carbon dioxide dissolving in the oceans, and the pH of the oceans decreasing. The increasing acidity makes it difficult for shellfish to make their shells.
2 Nitrogen – raw material for fertiliser manufacture and to freeze food; oxygen – medical treatments; argon – double glazing; neon – display lighting.

C1 14–16 Levelled questions: Changes in the Earth and its atmosphere
Working to Grade E
1 Outwards, from centre – core, mantle, crust.
2 a crust and upper part of the mantle;
  b solid;
  c convection currents;
  d radioactive processes.
3

Gas	Percentage
Nitrogen	80
Oxygen	20
Carbon dioxide, water vapour, noble gases	small proportion

4 algae; carbon dioxide; photosynthesis; oxygen.
5 a, d, e

Working to Grade C
6 a True
  b Over the past few years, the amount of carbon dioxide absorbed by the oceans has increased.
  c Most scientists agree that the increasing amounts of carbon dioxide in the atmosphere are causing global warming.
  d True
7 a The shapes of Africa and South America look as if they might once have fitted together; there are fossils of the same plants on both continents; there are rocks of the same type at the edges of the two continents where they might once have been joined.
  b Wegener was not a geologist; no one could work out how the continents might have moved.

8

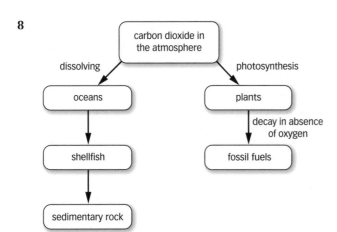

Working to Grade A*

9 Gases in the early atmosphere reacted with each other in the presence of sunlight, or lightning, to make the complex molecules that are the basis of life.

10 a Lightning
  b Two from: water vapour, carbon monoxide, hydrogen, hydrocarbon molecules and ammonia.
  c No one was around at the time to make observations.

11 a Fractional distillation.
  b Argon – double glazing; neon – display lighting.

C1 14–16 Examination questions: Changes in the Earth and its atmosphere

1 a The Earth's crust and the upper part of the mantle (1).
  b A few centimetres a year (1).
  c Radioactive process deep inside the Earth release heat (1). The heat causes convection currents within the mantle (1). These make the tectonic plates move.
  d Santiago is on a plate boundary, whereas Brasilia is not (1).
  e Earthquakes happen at plate boundaries when tectonic plates suddenly move relative to each other. No one can predict exactly when these movements will happen, so no one can predict exactly when earthquakes can happen (1).

2 a i

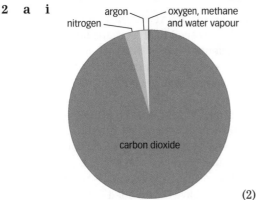

(2)

A pie chart is best, but bar chart is also acceptable.

ii Similarity – both atmospheres include nitrogen, carbon dioxide and argon gases (1). Difference – the percentage of carbon dioxide in the Martian atmosphere is much greater than the percentage of this gas in the atmosphere of the Earth (1). *There are several other acceptable differences such as the significantly lower percentages of oxygen and nitrogen in the Martian atmosphere.*

b i Erupting volcanoes (1).
  ii Carbon dioxide was removed by early plants and algae photosynthesising (1), and also by dissolving in the oceans (1).

c Humans have been burning fossil fuels in increasing amounts (1); humans also destroy forests, so less carbon dioxide is removed from the atmosphere by photosynthesis (1).

How Science Works: Polymers, plant oils, the Earth, and its atmosphere

1 The scientists working at an oil company and at the oilseed rape company might be biased because they are being paid by organisations that earn their money by selling one of the products under investigation.

2 The scientist's qualifications, or experience, or status within the scientific community.

3 Ethical arguments – Hari; Environmental arguments – Georgia, Krishnan; Economic arguments – Julia, Lydia; Social arguments – Imogen (and Hari).

4 Arguments that could be investigated scientifically – Georgia and Krishnan. Scientists can do investigations to collect evidence to help answer the questions.

C2 1: Ionic bonding

1

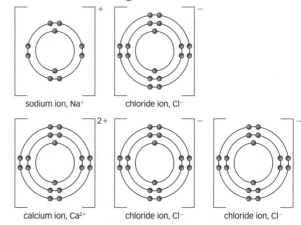

2 sodium + bromine → sodium bromide

3 There are strong electrostatic forces of attraction between oppositely charged sodium and chloride ions. These forces act in all directions. This is ionic bonding.

C2 2: Covalent bonding and metallic bonding

1 Simple molecules consist of a small number of atoms joined together by covalent bonds. Macromolecules are made up of large numbers of atoms joined together by covalent bonds to make a huge network.

2
H—H hydrogen

O=O oxygen

H—Cl hydrogen chloride

O—H water
|
H

3 In a giant covalent structure, the atoms are held together by strong covalent bonds to make a huge network. In a giant metallic structure, the electrons in the highest occupied level of the atoms are delocalised. These electrons move through the whole structure of the metal. The positive metal ions are arranged in a regular pattern. The whole structure is held together by strong electrostatic forces of attraction between the positive ions and the moving delocalised electrons.

C2 1–2 Levelled questions: Structure and bonding

Working to Grade E

1 Compounds – b, c, f, and g.

2 a True
 b When atoms share electrons, they form covalent bonds.
 c True
 d True
 e Elements in group 7 form ions with a charge of –1.

3 ions; positively; ions; negatively; noble gas; zero.

4 a Sodium chloride
 b Lithium oxide
 c Potassium iodide
 d Sodium bromide

5 Ionic – middle diagram; simple molecular – bottom diagram; giant covalent – top diagram.

Working to Grade C

6 Formulae that represent compounds are c and e.

7

Ion	Number of protons	Number of electrons
Li^+	3	2
F^-	9	10
Na^+	11	10
Cl^-	17	18
Mg^{2+}	12	10
Br^-	35	36
Ca^{2+}	20	18

8 a $[2.8]^+$
 b $[2.8.8]^-$
 c $[2.8]^{2+}$
 d $[2.8.8]^{2-}$
 e $[2.8.8]^{2+}$
 f $[2.8]^-$

9

Formula of positive ion	Formula of negative ion	Formula of compound
Na^+	Cl^-	NaCl
Mg^{2+}	O^{2-}	MgO
Ca^{2+}	Cl^-	$CaCl_2$
Rb^+	O^{2-}	Rb_2O

10 a H:H   H—H

 b :Cl:Cl: Cl—Cl

 c O::O   O=O

 d H×Cl: H—Cl

 e :O×H   O—H
    × |
    H   H

 f H×N×H   H—N—H
    × |
    H   H

 g   H   H
      × |
  H×C×H   H—C—H
      × |
      H   H

Working to Grade A*

11 In the structure, electrons from the highest energy level of each atom are delocalised. So there is a structure of positive metal ions in a regular pattern. The delocalised electrons move throughout the whole structure. Electrostatic forces of attraction between the positive ions and the delocalised electrons hold the structure together.

C2 1–2 Examination questions: Structure and bonding

1 a i negatively (1), chloride (1).
   ii 18 (1)
   iii

(2)

iv There are strong electrostatic forces of attraction between the oppositely charged ions. (1) These forces act in all directions in the lattice. (1) This type of bonding is called ionic bonding. (1)

b **i** Metallic bonding (1)

  ii Covalent (1)

  iii

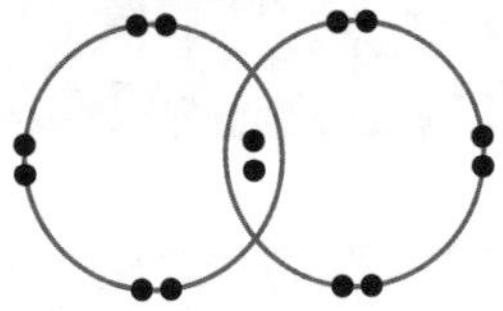

                                            (2)

C2 3: Inside molecules, metals and ionic compounds

1 Substances that consist of simple molecules do not conduct electricity because the molecules do not have an overall electric charge.

2

Type of bonding in substance	Melting and boiling points	Electrical conduction	Other properties
ionic	high	do not conduct when solid; conduct when molten and in solution	–
metallic	high	conduct	easy to bend
simple covalent	low	do not conduct	–

3 Substances that consist of simple molecules have relatively low boiling points because there are only weak intermolecular forces between the molecules. It is these forces that must be overcome when the substances melt or boil, not the much stronger covalent bonds between the atoms in the molecules.

C2 4: The structures of carbon, and nanoscience

1 Diamond is a form of the element carbon. In diamond, each carbon atom forms four covalent bonds with other carbon atoms to make a giant covalent structure.

2 The structure of diamond described in question 1 means that it is very hard. Graphite has a different structure. It is made up of layers. Within each layer, each carbon atom is joined to three other carbon atoms by strong covalent bonds. There are no covalent bonds between the layers, so the layers can slide over each other. This makes graphite soft and slippery.

3 Nanoparticles may be used in new computers, new catalysts, new coatings, highly selective sensors, stronger and lighter construction materials, and new cosmetics.

C2 5: Polymer structures, properties, and uses

1

Type of polymer	Properties	Structure
Thermosoftening	Soften easily when warmed. Can easily be moulded into new shapes.	Consist of individual polymer chains, with weak forces of attraction between the chains.
Thermosetting	Do not melt when heated.	Consist of polymer chains with cross-links between them.

C2 3–5 Levelled questions: Structure, properties, and uses

Working to Grade E

1 low; does not; do not have

2 **a** The forces between the oppositely changes ions in ionic compounds are strong.

  b True

  c Ionic compounds conduct electricity when molten and when dissolved in water.

  d Ionic compounds do conduct electricity when dissolved in water.

  e True

3 atoms, bendy, harder, less

4 **a** Nitinol

  b Dental braces

Working to Grade C

5 **a** J and N

  b L

  c M

  d M

  e K and M

  f L

6 **a** Thermosoftening polymers can be recycled because the forces of attraction between their particles are relatively weak, so they melt when heated and can be moulded into new shapes.

  b Thermosetting polymers cannot be recycled because the cross-links between their particles mean that they do not melt on heating. This means they cannot be moulded into new shapes.

  c Low density poly(ethene) and high density poly (ethene) are made under different conditions, and using different catalysts. This gives the two polymers their different properties.

7 The diagram shows that in diamond each carbon atom is joined to four others by strong covalent bonds to make a giant covalent structure. This makes diamond very hard. In graphite, each carbon atom is joined to only three carbon atoms by covalent bonds, to form layers of carbon atoms. The forces of attraction between the layers are very weak, so graphite is soft and slippery.

Working to Grade A*

8 a carbon, hexagonal
   b To deliver drugs to specific targets in the body; as catalysts; as lubricants; to reinforce materials such as graphite tennis racquets.

9 a B
   b G
   c B
   d M
   e B

10 In silicon dioxide, the atoms are held together by strong covalent bonds to form a giant structure. Large amounts of energy are required to overcome these strong covalent bonds, so silicon dioxide has high melting and boiling points. Nitrogen dioxide is made up of molecules. The forces of attraction between the molecules – the intermolecular forces – are weak compared to covalent bonds. It is these bonds which must be overcome when nitrogen dioxide melts or boils, so nitrogen dioxide has relatively low melting and boiling points.

11 Diamond has no delocalised electrons – or other particles with an overall electrical charge – so it cannot conduct electricity. Graphite has delocalised electrons. These particles are free to move so graphite can conduct electricity.

12 There are weak intermolecular forces between the polymer molecules in thermosoftening polymers. It is these forces which must be overcome when thermosoftening polymers melt, so they have low melting points. Thermosetting polymers have strong cross-links between the polymer molecules. These cross-links prevent thermosetting polymers melting when they are heated.

C2 3–5 Examination questions: Structure, properties, and uses

1 a stronger (1); stiffer (1).
   b i 1 (1)
      ii covalent (1)
      iii thermosoftening (1)
2 a 1 – 10 nm (1)
   b Carbon nanotubes are suitable for reinforcing the materials used to make wind turbines because they are very strong when subjected to pulling forces, (1) and very stiff. (1)
   c i Carbon nanotubes may cause cell death, (1) and may cause lung problems. (1)
      ii The tests on human cells were carried out on human cells outside the body and the studies on lung health were carried out on mice and rats (1). This means it is not possible to certain whether similar problems would be caused by nanotubes inside the human body. (1)

3 a 1 mark for each of the following points, up to a maximum of 4:
      - In pure platinum, the atoms are arranged in layers.
      - The layers can slide over each other very easily, making platinum relatively soft.
      - In the alloy, the different sized atoms of rhodium distort the structure of platinum, making it more difficult for them to slide over each other.
      - This makes the platinum-rhodium alloy harder than pure platinum.
   b Rhodium is a good conductor of electricity because it has delocalised electrons to carry the current. (2)

C2 6: Atomic structure

1 Nitrogen: atomic number = 7 and mass number = 14; iron: atomic number = 26 and mass number = 56; bromine: atomic number = 35 and mass number = 80
2 Atoms of the same element can have different numbers of neutrons, and so different mass numbers. Atoms of an element which have different numbers of neutrons are called isotopes.
3 $207 + \{(14 + [16 \times 3]) \times 2\} = 331$

C2 7: Quantitative chemistry

1 $(16 \div 40) \times 100 = 40\%$
2 The maximum theoretical yield may not be obtained because some of the product may be lost when it is separated from the reaction mixture; a reactant might have reacted in an unexpected way; the reaction might be reversible.
3

	Hydrogen	Carbon
Mass of each element, in g	0.4	1.2
A_r from periodic table	1	12
Mass divided by A_r	0.4	0.1
Simplest ratio	4	1
Formula	CH_4	

C2 8: Analysing substances

1 Advantages of instrumental analysis methods include their sensitivity, accuracy, and speed.
2 The sample is heated so that it becomes a mixture of vapours. Then a carrier gas is mixed with the mixture. The carries gas takes the mixture of vapours through a column packed with solid materials. Different substances in the vapour mixture travel through the column at different speeds, and become separated.

C2 6–8 Levelled questions: Atomic structure, analysis, and quantitative chemistry

Working to Grade E

1 Proton = 1; neutron = 1; electron = very small

2 Advantages – **a**, **c**, **d**

3 B, G, D, A, E, C, F

4 **a** True

   b The relative formula mass of a substance, in grams, is called one mole of that substance.

   c True

Working to Grade C

5 **a** Argon: atomic number = 18 and mass number = 40;

   b manganese: atomic number = 25 and mass number = 55;

   c zinc: atomic number – 30 and mass number = 65

6 **a** $(12 \times 8) + (1 \times 9) + 14 + (16 \times 2) = 151\,g$

   b $(12 \times 9) + (1 \times 8) + (16 \times 4) = 180\,g$

7 **a** $[39 \div (39 + 12 + 14)] \times 100 = 60\%$

   b $\{(7 \times 2) \div [(7 \times 2) + 12 + (16 \times 3)]\} \times 100 = 19\%$

   c $\{(14 \times 2) \div [(12 \times 10) + (1 \times 12) + (14 \times 2) + 16]\} \times 100 = 16\%$

8 **a** 4

   b D

   c A

9 The maximum theoretical yield may not be obtained because some of the product may be lost when it is separated from the reaction mixture; a reactant might have reacted in an unexpected way; the reaction might be reversible.

Working to Grade A*

10 58

11 The relative atomic mass of an element compares the mass of atoms of the element with the ^{12}C isotope. It is an average value for the isotopes of the element.

12 **a**

	Sulfur	Oxygen
Mass of each element, in g	3.2	3.2
A_r from periodic table	32	16
Mass divided by A_r	0.1	0.2
Simplest ratio	1	2
Formula		SO_2

b

	Carbon	Hydrogen
Mass of each element, in g	2.4	0.4
A_r from periodic table	12	1
Mass divided by A_r	0.2	0.4
Simplest ratio	2	4
Formula		C_2H_4

c

	Sodium	Nitrogen	Oxygen
Mass of each element, in g	2.3	1.4	4.8
A_r from periodic table	23	14	16
Mass divided by A_r	0.1	0.1	0.3
Simplest ratio	1	1	3
Formula		$NaNO_3$	

13 One mole of calcium carbonate has a mass of 100 g. So 10 g of calcium carbonate is 0.1 mole. The equation shows that one mole of calcium carbonate decomposes to make one mole of carbon dioxide. So 0.1 mole of calcium carbonate makes 0.1 mole of carbon dioxide. The mass of one mole of carbon dioxide is 44 g. So the maximum theoretical yield is 4.4 g.

14 Percentage yield = $(1.0 \div 1.5) \times 100 = 67\%$

C2 6–8 Examination questions: Atomic structure, analysis, and quantitative chemistry

1 **a** **i** 91 (1)

      ii 40 (1)

   b **i** First atom – number of neutrons = 52; (1) second atom – number of neutrons = 54. (1)

      ii isotopes (1)

2 **a** 1 mark is awarded for your calculation and 1 mark for the correct answer.

      $(12 \times 17) + (1 \times 19) + 14 + (16 \times 3) = 285$

   b **i** 369 g (1)

      ii 1 mark is awarded for your calculation and 1 mark for the correct answer.

      $[(16 \times 5) \div 369] \times 100 = 22\%$

   c **i** Morphine travels more slowly – it has the longer retention time. (1)

      ii The mixture is separated. (1)

3 **a** 1 mark is awarded for your calculation and 1 mark for the correct answer.

      $(1.5 \div 2) \times 100 = 75\%$

   b Some of the magnesium may have reacted with nitrogen from the air, so producing magnesium nitride as well as magnesium oxide. (1)

How science works: Structures, properties, and uses

1 Answer should include advantages and disadvantages such as the following:

   Nitinol stents have the advantage of changing shape to match the shape of the blood vessels they are holding open, but stainless steel stents do not. Nitinol also has the advantage of changing shape after being squashed, whereas stainless steel does not. A third advantage of nitinol stents is that blood clots are less likely to form on them than on stainless steel stents. It is possible, but unlikely, that nickel compounds can get into the blood of a person with a nitinol stent. This may increase the risk of cancer. Stainless steel stents contain small amounts of other metals, which may cause blockages in the blood vessel to form near the stent. Answer should then include a reasoned decision stating which type of stent is better.

2 The hospital might compare the costs of buying and inserting the two different types of stents.

3 a Anita Smith is funding herself, but Rachel Hooper is funded by a company that makes stainless steel stents, so might be influenced by the views of the company.

b Any well-reasoned answer can be given credit, for example: Professor Nadeem Hanif because he has a great deal of experience, and is not funded by an organisation that might bias his views.

C2 9: Rates of reaction and temperature

1 The graph shows that as the temperature increases, the reaction rate also increases.

2 Average rate = 35 cm³ ÷ 2 min = 17.5 cm³/min

3 The reaction has finished, so no more hydrogen is made.

C2 10: Speeding up reactions: concentration, surface area, and catalysts

1 Factors that affect reaction rate – temperature; concentration of reactants in solution; surface area of solid reactants; pressure for reactions involving a gaseous reaction.

2 The bigger the surface area of a solid reactant, the greater the number of reactant particles that are exposed. The greater the number of particles that are exposed, the greater the frequency of collisions and so the faster the reaction.

3 Catalysts speed up reactions without themselves being used up. They are important in industry because they make reactions fast enough to be profitable.

C2 9–10 Levelled questions: Rates of reaction

Working to Grade E

1 a and c increase the rate of reaction.

2 The minimum energy the particles must have to react.

3 A catalyst is a substance that speeds up a reaction without itself being used up in the reaction.

4 a The volume of gas increases quickly at first, and then more slowly.

   b i Rate = 50 cm³ ÷ 1 min = 50 cm³/min

   ii Rate = (70 – 50) cm³ ÷ 1 min = 20 cm³/min

c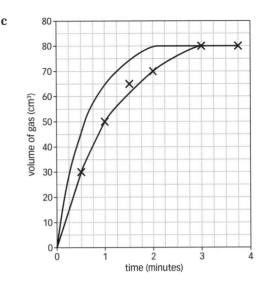

Working to Grade C

5 a Increasing the temperature increases the speed of the reacting particles so that they collide more frequently and more energetically. This increases the rate of reaction.

b Increasing the concentration of reactants in solutions increases the frequency of collisions and so increases the rate of reaction.

6 Using a catalyst helps to make the reaction fast enough to be profitable.

7 a Temperature

b Concentrations of solutions; volumes of solutions; how much the mixture is stirred or agitated.

c Range = 26 s to 400 s

d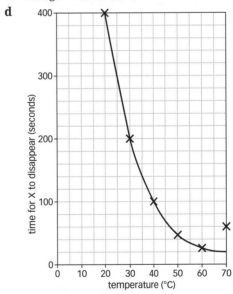

e The time for the cross to disappear at 70 °C is the anomalous result.

f The graph shows that the greater the temperature, the shorter the time for the cross to disappear and so the faster the reaction.

8 a Curve C

b The same amounts of reactants have been used each time.

C2 9–10 Examination questions: Rates of reaction

1 a Two from following list. 1 mark is awarded for each, up to a maximum of 2:
acid concentration, acid volume, amount of magnesium, size of pieces of magnesium.

   b i

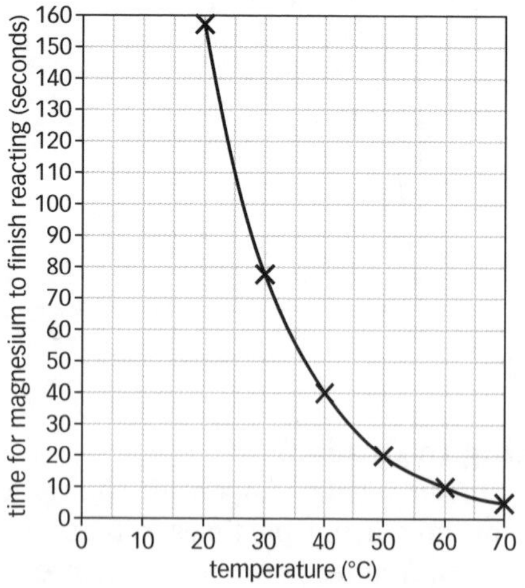

(1)

   ii 30 s (1)
   iii The graph shows that as the temperature increases, the time for the magnesium ribbon to disappear decreases, showing that the rate has increased. (2)
   iv Increasing the temperature increases the speed of the reacting particles so they collide more frequently and more energetically. (1) This increases the rate of reaction. (1)

C2 11: Exothermic and endothermic reactions

1 An exothermic reaction is one that transfers energy to the surroundings.
2 Endothermic reactions can be useful in sports injury packs, to cool damaged muscles.
3 The reaction gives out energy. At first, this energy heats up the reaction mixture. Then heat is transferred from the mixture to the surroundings, and the mixture cools to room temperature.

C2 11: Levelled questions: Exothermic and endothermic reactions

Working to Grade E

1 exothermic, endothermic, always.
2 Endothermic – sports injury packs; exothermic – hand warmers and self-heating cans for coffee.
3 d – Thermal decomposition
4 a $(14 + 9 + 10) \div 3 = 11\,°C$
   b Anomalous result – test 2 with nitric acid.
   c To spot any anomalous data;
      To improve the accuracy of the data.
   d The volume of the acid and the concentration of the acid.

Working to Grade C

5 Exothermic reactions – a, c, d
6 a A, B, D
   b C, E
   c C, E
   d A, B, D
7 From left to right, the reaction is endothermic and from right to left the reaction is exothermic.

C2 11: Examination questions: Exothermic and endothermic reactions

1 a After the reaction, the temperature of the solution was higher than the temperatures of the reactant solutions before the reaction. This shows that the reaction is exothermic. (1)
   b Oxidation and combustion. (1)
   c 1 mark for one answer from: hand warmers, self-heating coffee cans.
   d i At first, the temperature decreases. (1)
      ii The sports injury pack takes in heat energy from the surroundings and returns to the temperature of the surroundings. (1)

C2 12: Acids and bases

1 Bases – copper oxide, zinc oxide, magnesium oxide, sodium hydroxide, potassium hydroxide (or any other metal oxides or hydroxides); alkalis – sodium hydroxide, potassium hydroxide, (or any other soluble metal oxides or hydroxides).
2 Ammonia dissolves in water to make ammonium ions (NH_4^+) and hydroxide ions (OH^-). The hydroxide ions make the solution alkaline.
3 hydrochloric acid + potassium hydroxide → potassium chloride + water
4 $HCl\,(aq) + KOH\,(aq) \rightarrow KCl\,(aq) + H_2O\,(l)$

C2 13: Making salts

1 A salt is a compound that contains metal or ammonium ions. Salts can be made from acids.
2 Add small pieces of zinc to sulfuric acid until there is no more bubbling and a little zinc remains unreacted. Filter to remove the unreacted zinc. Heat the solution over a water bath, until about half its water has evaporated. Leave the solution to stand for a few days to form zinc sulfate crystals.
3 Zinc chloride.
zinc oxide + hydrochloric acid → zinc chloride + water
4 Place some nitric acid in a conical flask. Add a few drops of Universal indicator. Add potassium hydroxide until the solution is just neutral. Add one spatula measure of charcoal powder to remove the colour of the Universal indicator. Filter. Pour the filtrate into an evaporating basin. Heat over a beaker of boiling water until about half the water of the

solution has evaporated. Leave the solution to stand for a few days to form potassium nitrate crystals.

$$\text{nitric acid} + \text{potassium hydroxide} \rightarrow \text{potassium nitrate} + \text{water}$$

C2 14: Precipitation and insoluble salts

1   A precipitation reaction is one in which two solutions react to form a precipitate.
2   Barium chloride and sodium sulfate (or any other soluble compound of barium and any soluble sulfate).

C2 12–14 Levelled questions: Acids, bases, and salts

Working to Grade E
1   (aq) – dissolved in water; (s) – solid; (l) – liquid; (g) – gas
2   **a**   chlorides
    b   nitrates
    c   sulfates
3   alkaline, ammonium, fertilisers
4   **a**   True
    b   Bases are oxides of metals
    c   Sulfur dioxide is not a base
    d   True
    e   True
    f   True
    g   True
    h   True
    i   Hydroxide ions make solutions alkaline or hydrogen ions make solutions acidic
    j   True
    k   A solution with a pH of 6 is acidic
5   C, A, E, B, F, D

Working to Grade C
6

Solution type	pH
Acidic	less than 7
Neutral	7
Alkaline	more than 7

7   **a**   Magnesium chloride
    b   Copper sulfate
    c   Potassium nitrate
    d   Magnesium sulfate
8   $H^+ (aq) + OH^- (aq) \rightarrow H_2O (l)$
9

Solution contains...	acidic, alkaline, or neutral?
An equal number of OH⁻ and H⁺ ions.	neutral
More H⁺ ions than OH⁻ ions.	acidic
More OH⁻ ions than H⁺ ions.	alkaline

10 Add small amounts of magnesium oxide to sulfuric acid until a little solid magnesium oxide remains unreacted. Filter to remove the unreacted magnesium oxide. Heat the solution over a water bath, until about half its water has evaporated. Leave the solution to stand for a few days to form magnesium sulfate crystals.
11 Place some hydrochloric acid in a conical flask. Add a few drops of Universal indicator. Add sodium hydroxide until the solution is just neutral. Add one spatula measure of charcoal powder to remove the colour of the Universal indicator. Filter. Pour the filtrate into an evaporating basin. Heat over a beaker of boiling water until about half the water of the solution has evaporated. Leave the solution to stand for a few days to form sodium chloride crystals.
12 nitrate, potassium, precipitate, lead iodide, potassium nitrate
13 **a**   Lead iodide
    b   Barium sulfate
    c   Lead iodide
    d   Silver hydroxide
    e   Barium sulfate
14 The pairs of solutions given below are examples only; other combinations will also produce the named insoluble salts.
    a   Lead nitrate and sodium chloride
    b   Calcium nitrate and sodium sulfate
    c   Lead nitrate and sodium sulfate
    d   Barium chloride and sodium sulfate

C2 12–14 Examination questions: Acids, bases, and salts

1   **a**   Hydrochloric acid (1)
    b   To remove unreacted zinc metal. (1)
    c   Crystallisation (1)
    d   Two from following list. 1 mark awarded for each up to a maximum of 2:
        do not touch the product; wash hands after the practical; wear eye protection; work in a well-ventilated laboratory; follow instructions for the disposal of chemicals.
2   **a**   **i**   Copper oxide, magnesium oxide, sodium hydroxide (1)
             ii Sodium hydroxide (1)
    b   Name: hydroxide; (1) formula: OH⁻ (1)
3   **a**   1 mark for each of: lead nitrate and potassium iodide solutions
    b   Mix the two solutions. Filter. The solid lead iodide remains in the filter paper. (2)
    c $Pb^{2+}(aq) + 2 I^- (aq) \rightarrow PbI_2 (s)$
        1 mark awarded for each correct state symbol.

C2 15: Electrolysis 1

1   Positive electrode – chlorine; negative electrode – copper.
2   Positive electrode – oxygen; negative electrode – hydrogen.

3 Positive electrode: $Cl^- \rightarrow Cl + e^-$
then $Cl + Cl \rightarrow Cl_2$
Negative electrode: $Cu^{2+} + 2e^- \rightarrow Cu$

C2 16: Electrolysis 2

1 To protect a metal object from corrosion by coating it with an unreactive metal that does not easily corrode; to make an object look attractive.

2 Aluminium and carbon dioxide.

3 Positive electrode – chlorine gas; negative electrode – hydrogen gas.

C2 15–16 Levelled questions: Electrolysis

Working to Grade E

1 **a** and **b**

2 melted/dissolved; melted/dissolved; ions; solution

3 electrolyte – a liquid or solution that is broken down when electricity passes through it; electrolysis – the process by which electricity breaks down a liquid or solution; electroplating – covering an object with a layer of metal in an electrolysis cell; electrodes – piece of metal or graphite through which electricity enters or leaves an electrolysis cell

4 bromide; positively; positive; negative

Working to Grade C

5 **a** At the negative electrode, positively charged ions gain electrons.

b True

c If an ion gains electrons, the ion is reduced.

d True

6

Solution	Positive electrode	Negative electrode
copper chloride	chlorine	copper
potassium bromide	bromine	hydrogen
silver nitrate	oxygen	silver
magnesium nitrate	oxygen	hydrogen
copper carbonate	oxygen	copper
sodium sulfate	oxygen	hydrogen

7 Positive electrode – chlorine gas, used for making bleach, plastics, and sterilising water; negative electrode – hydrogen gas, used for making margarine and ammonia; solution of sodium hydroxide formed in electrolyte vessel, used for making soap.

8
- Positive electrode **4**
- Positive ions gain electrons at this electrode **1**
- Negative electrode **1**
- Electrolyte of liquid aluminium and cryolite **2**
- Liquid aluminium forms at this electrode **3**
- Oxide ions are oxidised at this electrode **4**
- Carbon dioxide gas forms here **4**
- This electrode is made of carbon **4**
- Reduction happens at this electrode **1**
- Negative ions are attracted to this electrode **4**

Working to Grade A*

9 **a** Negative electrode: $Pb^{2+} + 2e^- \rightarrow Pb$
Positive electrode: $Br^- \rightarrow Br + e^-$
then $Br + Br \rightarrow Br_2$

b Negative electrode: $Cu^{2+} + 2e^- \rightarrow Cu$
Positive electrode: $Cl^- \rightarrow Cl + e^-$
then $Cl + Cl \rightarrow Cl_2$

c Negative electrode: $Al^{3+} + 3e^- \rightarrow Al$
Positive electrode: $O^{2-} \rightarrow O + 2e^-$
then $O + O \rightarrow O_2$

C2 15–16 Examination questions: Electrolysis

1 **a** The rhodium is harder, so protects the silver from scratches; (1) the rhodium does not react with gases from the air, so it protects the silver from corrosion. (1)

b **i** So that positive rhodium ions are attracted towards it, and then gain electrons to form rhodium metal on the surface of the ring. (1)

ii Reduction (1)

iii $Rh^{3+} + 3e^- \rightarrow Rh$ (1)

How Science Works: Rates, energy, salts, and electrolysis

1 **a** Dependent variable – time to collect $50\,cm^3$ of hydrogen gas; independent variable – temperature; control variables – (three from) mass of magnesium, surface area of magnesium, concentration of acid, volume of acid, stirring.

b 48

c Graph with temperature on x-axis and time to collect $50\,cm^3$ hydrogen gas on y-axis. A smooth curve should be drawn.

d The results do support the hypothesis as the rate of reaction increases with increasing temperature.

2 **a** A new catalyst does not have to be bought each time – this reduces costs.

b The catalyst does not have to be disposed of; smaller amounts of catalyst are required, so reducing the impact on the environment of producing the catalyst material.

C3 1: The periodic table

1 Newlands listed the elements then known in order of atomic weight. Every eighth element had similar properties. He used this pattern to group the elements. He called his idea the 'law of octaves.'

2 Mendeleev swapped the positions of some pairs of elements so that they were grouped with elements with similar properties; Mendeleev left gaps for elements which he predicted did exist, but which had not then been discovered.

3 Strong, hard, high density, high melting point, form ions with different charges, form coloured compounds, react slowly or not at all with water and oxygen.

C3 2: Group 1 – The alkali metals

1 Three rows from:

	Group 1 elements	Transition elements
Density	low	high
Hardness	soft	hard
Strength	strong	weak
Melting point	low	high (except mercury)
Compound colour	white	coloured
Reaction with oxygen	vigorous	not vigorous, or does not occur
Reaction with water	vigorous	not vigorous, or does not occur

2 Similarities: Both lithium and potassium react with water to form a hydroxide and hydrogen gas. The reactions are vigorous, and the hydrogen gas propels the metal around on the surface of the water as it reacts.
Differences: a lilac flame is seen when potassium reacts with water. There is no flame when lithium reacts with water.

3 The trend can be explained by the energy level of the outer electrons. Potassium is lower down Group 1 than lithium. The outermost electron of potassium is in a higher energy level than that of lithium. This means that, in reactions, potassium gives away its outermost electron more easily than lithium. Potassium is more reactive than lithium.

C3 3: Group 7 – The halogens

1 Going down the group, the melting points and boiling points increase.

2 Iron and chlorine have the more vigorous reaction, because chlorine is more reactive than iodine.

3 **a**

chlorine +	sodium iodide	$\rightarrow$	sodium chloride	+ iodine
Cl_2 +	$2NaI$	$\rightarrow$	$2NaCl$	+ I_2

b

bromine +	potassium iodide	$\rightarrow$	potassium bromide	+ iodine
Br_2 +	$2KI$	$\rightarrow$	$2KBr$	+ I_2

C3 1–3 Levelled questions: The periodic table

Working to Grade E

1 transition; different; coloured; catalysts

2 **a** 1
   b 3
   c 7

3 **a** In the periodic table, a vertical column is called a group.
   b True
   c Mendeleev left gaps in his periodic table for elements he predicted did exist, but had not yet been discovered.

4 **a** manganese
   b potassium
   c iron
   d manganese

Working to Grade C

5

Name of compound	Formula of metal ion in compound	Appearance and state of compound at room temperature
potassium chloride	K+	white solid
sodium bromide	Na+	white solid
lithium chloride	Li+	white solid

6 **a** sodium + chlorine → sodium chloride
   b lithium + oxygen → lithium oxide
   c sodium + water → sodium hydroxide + hydrogen
   d potassium + water → potassium hydroxide + hydrogen

7 Equations for the pairs that react:
   a chlorine + potassium bromide → potassium chloride + bromine
   b No reaction
   c bromine + potassium iodide → potassium bromide + iodine
   d No reaction
   e chlorine + potassium iodide → potassium chloride + iodine

8 **a** 1
   b 7
   c 1
   d 7
   e 1
   f 1

Working to Grade A*

9 **a** $2Na(s) + Br_2(l) \rightarrow 2NaBr(s)$
   b $2Li(s) + 2H_2O(l) \rightarrow 2LiOH(aq) + H_2(g)$
   c $Cl_2 + 2KI \rightarrow 2KCl + I_2$

10 When a metal reacts with a halogen, the metal atoms give each halogen atom **one** extra electron. This completes the **outer** energy level of the halogen atom. The closer the outer energy level is to the nucleus, the **greater** the attraction between the newly-added electrons and the nucleus. So the lower down Group 7 an electron is, the **less** easily its atoms gain electrons, and the less reactive the element is.

C3 1–3 Examination questions: The periodic table

1   a   i   Mendeleev used the atomic weights published by Cannizzaro to arrange the elements in order. (1)

       ii   Mendeleev's predictions, of the positions of the elements in the periodic table that had not then been discovered, were correct. (1)

     b   i   1 mark awarded for either atomic number/proton number.

       ii   3 marks available: 1 mark will be awarded for your explanation, 1 mark for including examples and 1 mark for providing at least one correct electronic structure of an element. For example:
           All the elements in a group have the same number of electrons in their highest occupied energy level. The number of electrons in the highest occupied energy level is the same as the group number, for the main groups. For example the electronic structure of sodium (a member of Group 1) is 2.8.1. This shows it has 1 electron in its highest occupied energy level.

2   **Here is a guide to the marking for an essay-style question:**
   To gain 5/6 marks – All information in the answer must be relevant, clear, organised, and presented in a structured and coherent format. Specialist terms should be used appropriately. Few, if any, errors in grammar, punctuation, and spelling. Answer should include 5 or 6 points from those below.
   To gain 3/4 marks – Most of the information should be relevant and presented in a structured and coherent format. Specialist terms should usually be used correctly. There may be occasional errors in grammar, punctuation, and spelling. Answer should include 3 or 4 points from those below.
   To gain 1/2 marks – Answer will be simplistic. There may be limited use of specialist terms. Errors of grammar, punctuation, and spelling will prevent communication of the science. Answer should include 1 or 2 points of those listed below.
   Points to include:
   - Transition elements (except mercury) have higher melting points than the Group 1 elements.
   - Transition elements have lower densities.
   - Transition elements are stronger.
   - Transition elements are harder.
   - Transition elements form coloured compounds, Group 1 elements form white compounds.
   - Transition elements form ions with different charges, Group 1 elements form ions with a charge of +1.
   - The Group 1 elements react vigorously with oxygen and water, the Transition elements react less vigorously, if at all.

3   a   Gas (1)
     b   The boiling point increases. (1)
     c   hydrogen + fluorine → hydrogen fluoride (1)
     d   i   Iodine (1)
       ii   2 marks available. 1 mark will be awarded for naming the product and 1 mark for explaining how they will react. For example:
           The element will react with hydrogen to form hydrogen iodide. The reaction will be less vigorous than the reaction of hydrogen with bromine.
     e   i   A maximum of 4 marks are available. 1 mark will be awarded for describing the trend in reactivity. Then 1 mark per reaction observation will be awarded up to a maximum of 2. 1 mark per displacement reaction observation will be awarded up to a maximum of 2. For example:
           Going down the group, the elements get less reactive. The reactions of the halogens with hydrogen illustrate this trend – the reaction with fluorine happens spontaneously and explosively; the reaction of hydrogen with bromine requires a lighted splint to get it started. In group 7, a more reactive halogen displaces a less reactive halogen from an aqueous solution of its salt. For example, chlorine displaces iodine from a solution of sodium iodide because chlorine is more reactive than iodine. Iodine does not displace chlorine from sodium chloride solution because iodine is les reactive than chlorine.
       ii   This trend can be explained by the energy levels of the outer electrons in halogen atoms. (1) For example, when hydrogen reacts with chlorine, hydrogen atoms give each chlorine atom one extra electron. The electron completes the outer energy level of the chlorine atom, forming a chloride ion. The higher the energy level of the outer electrons, the less easily electrons are gained. (1)

C3 4: Making water safe to drink

1   A suitable source is chosen; the water is filtered; the water is sterilised by adding chlorine.

2   Advantages – prevents tooth decay; less money spent treating dental problems.
   Disadvantages – expensive; would be unnecessary if everyone looked after their teeth properly; very large amounts of fluoride compounds can make teeth yellow.
   A conclusion for or against adding fluorine to water should then be added, and backed up by a reason.

3

Type of water filter	What it removes
carbon	chlorine and other molecules with unpleasant tastes and smells
silver	bacteria
ion exchange resin	cadmium, lead, copper ions (and calcium and magnesium ions)

C3 5: Hard and soft water

1 Dissolved calcium and magnesium ions make water hard.

2 Hard water advantages – calcium ions prevent heart disease and help the development and maintenance of teeth and bones.
Hard water disadvantages – forms scale in kettles and boilers, increasing energy bills and maybe increasing emission of greenhouse gases; forms scum with soap, so increasing amount of soap needed.
Soft water advantages – does not form scale or scum.
Soft water disadvantages – no calcium ions for maintenance of teeth and bones.

3 Boiling – this removes temporary hardness; adding sodium carbonate – this removes both types of hardness; passing water through an ion exchange resin – the calcium and magnesium ions are replaced by sodium, or hydrogen, or potassium ions.

C3 4–5 Levelled questions: Water

Working to Grade E

1 **a**, **c**, **d**, **e** = H; **b** = S

2 Choose an appropriate source – to minimise the treatment needed to make the water safe to drink;
Pass the water through filter beds – to remove solids from the water;
Add chlorine – to kill bacteria in the water.

3 lather; scum; more; detergents

4 Compounds that may be dissolved in hard water: **a, c, d**

Working to Grade C

5 Arguments for: prevents tooth decay, reduces dental treatment bills.
Arguments against: expensive, unnecessary for those who look after their teeth properly.

6 Permanent hardness is not removed on boiling; temporary hardness is removed on boiling.

7 **a** Anomalous result – village C, run 3
b A – 3; B – 20; C – 44
c So that she can calculate a mean from the three results so increasing the accuracy of her data **or** to identify anomalous results.
d A, B, C
e A
f B and C

8 1 – D; 2 – A; 3 – F and B; 4 – C; 5 – E

9 Sodium carbonate is soluble in water. In hard water, its carbonate ions react with dissolved calcium and magnesium ions. Calcium carbonate and magnesium carbonate form as precipitates. They can be removed by filtering.

10 The energy to evaporate the water is not supplied from the Sun, but is supplied from fossil fuel sources, for example.

Working to Grade A*

11 D, A, E, C, F, B

C3 4–5 Examination questions: Water

1 **Here is a guide to the marking for an essay-style question:**
To gain 5/6 marks – All information in the answer should be relevant, clear, organised, and presented in a structured and coherent format. Specialist terms will be used appropriately. Few, if any, errors in grammar, punctuation, and spelling. Answer should include 5 or 6 points from those below.
To gain 3/4 marks – Most of the information should be relevant and presented in a structured and coherent format. Specialist terms will usually be used correctly. There may be occasional errors in grammar, punctuation, and spelling. Answer should include 3 or 4 points of those listed below.
To gain 1/2 marks – Answer will be simplistic. There may be limited use of specialist terms. Errors of grammar, punctuation, and spelling will prevent communication of the science. Answer should include 1 or 2 points of those listed below.
Points to include:
- Chlorine kills bacteria in water from source to tap.
- Adding chlorine to water prevents illness and death from waterborne diseases.
- Adding chlorine to water reduces health bills.
- Some people do not like the taste and smell of chlorine in tap water.
- Adding chlorine to tap water means that companies can make money selling water filters that remove it from water.
- Adding chlorine to water increases the costs of water treatment, compared to not adding it.
- Conclusion, backed up by reasons.

2 **a** The student has a water softener at home, but it doesn't work when the column is saturated with calcium ions (1);
On different days, the water company supplies water from different sources (1).
b 1 mark will be awarded for each of the following points, up to a maximum of 3:
- Add soap solution from the burette to each water sample in turn.
- Record the volume of soap solution required to form permanent lather.
- The smaller the volume required, the softer the water.

c **i** 2 October (1)

    ii Take more than one sample on each date, find the volumes of soap solution required for each one, and calculate the mean value added. (1)

    iii 2 October (1)

3 **a** **i** calcium (1) and magnesium (1).

    ii The water in beaker A has more Ca^{2+} and Mg^{2+} ions in it. (1)
The water in beaker B has more Na^+ ions in it. (1)

  b 3 marks available. 1 mark awarded for an advantage and 1 mark for a disadvantage, up to a total of 2.
Advantages – less scum with soap, and less scale in kettles, reducing bills for soap and energy.
Disadvantages – less calcium in water to help maintain healthy bones and teeth, less calcium to help protect against heart attacks.
Final mark awarded for a conclusion, with supporting evidence.

4 **a** Temporary hard water is softened on boiling, permanent hard water is not softened on boiling. (1)

  b Sodium carbonate is soluble in water. In hard water, its carbonate ions react with dissolved calcium and magnesium ions. (1) Calcium carbonate and magnesium carbonate form as precipitates. They can be removed by filtering. (1)

  c Boil the water. (1) Temporary hard water contains hydrogen carbonate ions (HCO_3^-). On heating, these ions decompose to produce carbonate ions (CO_3^{2-}). (1) The carbonate ions react with calcium ions (Ca^{2+}) or magnesium ions (Mg^{2+}) in the water to make calcium carbonate or magnesium carbonate. (1) Calcium carbonate and magnesium carbonate are insoluble in water, so they form as precipitates. The precipitates form scale in kettles and boilers. The boiled water contains few calcium or magnesium ions. It has been softened. (1)

C3 6: Calculating and explaining energy change

1 Joules, J

2 29 400 J

3 To help minimise heat losses to the surroundings.

C3 7: Energy-level diagrams

1

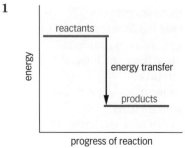

2 A catalyst provides a different pathway for a chemical reaction that has a lower activation energy.

3 The energy released in forming four new O–H bonds in the water molecules is greater than the energy needed to break the existing two H–H and one O=O bonds.

C3 8: Fuels

1 Hydrogen can be burned as a fuel in combustion engines, forming water.
It can be also used in fuel cells that produce electricity to power vehicles.

2 Carbon dioxide (a greenhouse gas) is produced; many hydrocarbon fuels are not renewable; burning hydrocarbons at high temperatures makes oxides of nitrogen, which destroy ozone in the upper atmosphere. **There are many other possible correct answers.**

C3 6–8 Levelled questions: Calculating and explaining energy change

Working to Grade E

1 **a** True

  b 1000 J = 1 kJ

  c In the equation $Q = mc\Delta T$, ΔT represents temperature change.

  d True

  e In a chemical reaction, energy is released as new bonds are formed.

  f In a chemical reaction, energy must be supplied to break bonds.

  g The new pathway has a lower activation energy.

2 **a** F

  b A

  c C

  d Exothermic – energy stored by products is less than energy stored by reactants.

Working to Grade C

3 **a** Hydrogen

  b Advantages – water is the only product; hydrogen can be obtained from renewable sources of methane.
Disadvantages – energy is required to produce hydrogen; few filling stations supply hydrogen gas.

4 14 700 J

5 **a** 2 940 J

  b Exothermic – temperature increased.

6 **a** **i** –350 kJ/mol

    ii +400 kJ/mol

  b Reaction 1 – energy stored in reactants is more than energy stored in products.

7

Reaction	Bonds that break	Bonds that are made
a	H–H Cl–Cl	H–Cl H–Cl
b	H–H H–H O=O	H–O H–O H–O H–O
c	C–H Cl–Cl	C–Cl H–Cl

Working to Grade A*

8 **a** O=O
   b Cl–Cl
   c O=O
9 **a** $-185\,kJ/mol$
   b $-483\,kJ/mol$
   c $-115\,kJ/mol$

C3 6–8 Examination questions: Calculating and explaining energy change

1 **a** Keep the cereal the same distance from the water being heated. (2)
   b $\dfrac{(32+36=38)}{3} = 35.3\,°C$ (1)
   c 1 mark will be awarded for showing your working, 1 mark for the correct answer.
      $100 \times 4.2 \times 35.3 = 14\,826\,J$
   d **i** During the student's experiment, much energy is transferred to the surroundings, the calorimeter, and the rest of the apparatus. This is not the case for the value on the packet. (1)
      ii Add a lid (or any other suitable suggestion). (1)
2 **a** **i** 1 mark is awarded for each impact identified, up to a maximum of 2. For example: less methane is released to the atmosphere; less cow manure is available to fertilise crops; less fossil fuel is used; or any other suitable answer.
      ii The farmer will make money from selling the gas. (1)
   b The energy stored by the reactants is greater than the energy stored by the products. (2)
   c Bonds broken:
      4 (C–H) and 2 (O=O) (1)
      Energy required to break these bonds
      $= (4 \times 413) + (2 \times 497)$
      $= 2646\,kJ/mol$ (1)
      Bonds made:
      2 (C=O) and 4 (H–O)
      Energy given out on forming these bonds
      $= (2 \times 743) + (4 \times 463)$
      $= 3338\,kJ/mol$ (1)
      Overall energy change for reaction
      = energy required to break bonds – energy given out on forming new bonds
      $= 2646 - 3338$
      $= -692\,kJ/mol$ (1)

How Science Works: The periodic table, water, and energy changes

1 **a** Town D
   b Town A, mean = 5; town B, mean = 28; town C, mean = 7; town D, mean = 58; town E, mean = 7
   c Bar chart, since one of the variables (town) is categoric.
   d Bar chart should include towns A to E labelled on the x-axis; number of drops, with an even scale on the y-axis; correctly plotted bars.
2 **a** Line graph should include volume of sodium carbonate added (cm^3) labelled on the x-axis; number of drops, with an even scale, on the y-axis; correctly plotted data; line of best fit.
   b The value for $7\,cm^3$ of sodium carbonate solution is anomalous.
   c The graph shows that as the volume of sodium carbonate solution added increases, the number of drops of soap solution required for a permanent lather decreases. This indicates that the greater the volume of sodium carbonate solution added, the softer the water becomes. Once $5\,cm^3$ of sodium carbonate solution have been added, the water does not become any softer. This suggests that all the calcium or magnesium ions that were dissolved in the water have now been removed.

C3 9: Identifying positive ions and carbonates

1 Copper hydroxide – blue; magnesium hydroxide – white; iron (II) hydroxide – green.
2 Dissolve the salt in pure water. Add a few drops of sodium hydroxide solution. If the salt contains aluminium ions, the precipitate will be white. Add excess sodium hydroxide solution to the precipitate. If the precipitate dissolves, the salt contains aluminium ions.
3 Add a few drops of dilute hydrochloric acid to the solid. If it fizzes, test the gas evolved with limewater. If the limewater goes cloudy, the salt was a carbonate.

C3 10: Identifying halides and sulfates, and doing titrations.

1 Dissolve a little of the solid to be tested in dilute nitric acid. Add silver nitrate solution. If a cream precipitate forms, the salt was a bromide.
2 To improve the accuracy, and so ensure the value calculated is as close to the true value as possible.
3 $0.20\,mol/dm^3$

C3 11: More on titrations

1 $0.05\,mol$
2 $0.0048\,mol$
3 $0.104\,mol/dm^3$

C3 9–11 Levelled questions: Further analysis and quantitative chemistry

Working to Grade E

1

Compound of...	Flame colour
lithium	crimson
sodium	yellow
potassium	lilac
calcium	red
barium	green

2

Name of precipitate	Colour
aluminium hydroxide	white
copper(II) hydroxide	blue
iron(II) hydroxide	green
calcium hydroxide	white
iron(III) hydroxide	brown
magnesium hydroxide	white

3 aluminium hydroxide

Working to Grade C

4 hydrochloric, barium chloride, white, nitric, silver nitrate, white.

5 Add a few drops of dilute hydrochloric acid to the solid, if it fizzes, test the gas with limewater. If the limewater turns cloudy, the solid is a carbonate.

6 a Use a **pipette** to measure out exactly 25.00 cm^3 of sodium hydroxide solution.

   b Transfer the solution to a **conical flask**.

   c Using a funnel, pour the hydrochloric acid into a burette. Add a few drops of indicator to the **conical** flask.

   d Read the scale on the burette. Add hydrochloric acid from the burette to the sodium hydroxide solution in the **conical flask** until the indicator changes colour.

   e Repeat steps (a) to (d) **until consistent results are obtained.**

7 a 9.70 cm^3

   b 9.77 cm^3

Working to Grade A*

8 copper sulfate 4 g/dm^3
   sodium chloride 30 g/dm^3
   magnesium sulfate 200 g/dm^3

9 0.025 mol

10 0.0047 mol

11 58.5 g

12 23.75 g

13 0.924 mol/dm^3

14 0.0536 mol/dm^3

C3 9–11 Examination questions: Further analysis and quantitative chemistry

1 a Dip the end of a clean nichrome wire into the salt. Hold the end of the nichrome wire in a hot Bunsen flame. Observe the flame colour. (1)

   b 3 marks available. 1 will be awarded for each correct alternative conclusion that is suggested and that is supported with evidence from the table. For example:
   The mixture contains calcium ions and lithium ions. Evidence – the flame colour could be caused by both lithium and calcium ions.
   The mixture contains calcium ions and magnesium ions. Evidence – the flame colour could be caused by calcium ions, and the white precipitate could be a mixture of calcium hydroxide and magnesium hydroxide formed when sodium hydroxide solution reacted with calcium and magnesium ions in solution.
   The mixture contains lithium ions and magnesium ions. Evidence – the flame colour could be caused by lithium ions, and the white precipitate could be magnesium hydroxide formed when sodium hydroxide solution reacted with magnesium ions in solution.

   c Test 4 – sulfuric acid has been added to the mixture, so the test will be positive for sulfate ions, whatever the salt. (1)

   d Carbonate ions – Test 3, fizzed with acid, and the gas produced made limewater cloudy; (1)
   Chloride ions – Test 5, white precipitate with silver nitrate solution. (1)

2 **Here is a guide to the marking for an essay-style question:**
   To gain 5/6 marks – All information in the answer should be relevant, clear, organised, and presented in a structured and coherent format. Specialist terms will be used appropriately. Few, if any, errors in grammar, punctuation, and spelling. Answer should include 5 or 6 points from those below.
   To gain 3/4 marks – Most of the information should be relevant and presented in a structured and coherent format. Specialist terms will usually be used correctly. There may be occasional errors in grammar, punctuation, and spelling. Answer should include 3 or 4 points of those listed below.
   To gain 1/2 marks – Answer will be simplistic. There may be limited use of specialist terms. Errors of grammar, punctuation, and spelling will prevent communication of the science. Answer should include 1 or 2 points of those listed below.
   Points to include:
   • Use a pipette to measure out 25 cm^3 of sodium hydroxide solution.

- Transfer the solution to a conical flask.
- Add a few drops of indicator to the solution in the flask.
- Pour hydrochloric acid into a burette and read the scale.
- Allow hydrochloric acid to run into the conical flask, with swirling.
- When the indicator changes colour, the end point has been reached.
- Repeat until several consistent results are obtained, but add acid one drop at a time as the end point is approached.

3 a i To find out the approximate volume of acid required. (1)
   ii To improve accuracy. (1)
   iii $(11.90 + 12.00 + 12.10) \div 3 = 12.00$ (1)
  b No – much more DCPIP (80 times) was required with the orange juice than the blackcurrant drink, indicating that the orange juice contained more vitamin C. (1)

4 $H_2SO_4 + 2NaOH \rightarrow Na_2SO_4 + H_2O$
  Mass of one mole of sulfuric acid
  $= (2 \times 1) + 32 + (16 \times 4) = 98\,g$
  Concentration of sulfuric acid solution
  $= 9.8\,g \div 98\,g = 0.1\,mol$ (1)
  Number of moles of sulfuric acid in $23.00\,cm^3$ of solution
  $= (23.00 \div 1000) \times 0.1$
  $= 0.0023\,moles$ (1)
  The equation shows that 1 mole of sulfuric acid reacts with 2 moles of sodium hydroxide. So in $25.00\,cm^3$ sodium hydroxide solution there are $0.0023 \times 2 = 0.0046$ moles of sodium hydroxide. (1)
  So in $1000\,cm^3$ of sodium hydroxide solution there are $0.0046 \times (1000 \div 25) = 0.184$ moles of sodium hydroxide.
  So the concentration of sodium hydroxide solution is $0.184\,mol/dm^3$ (1)

C3 12: Ammonia

1 Nitrogen – separated from the air; hydrogen – obtained from natural gas.
2 A temperature of $450\,°C$ is chosen for the Haber process. At this temperature, the percentage yield is smaller than it is at lower temperatures. But the rate of the reaction increases as temperature increases. At $450\,°C$ the yield and rate are both acceptable. The environmental impact and energy requirements would be less at lower temperatures, but the rate would be too slow.
3 A pressure of $200\,atm$ is chosen for the Haber process. The higher the pressure, the higher the yield. But at high pressures the equipment and operating costs increase. The pressure chosen is a compromise between maximising yield and minimising costs.

C3 13: Ammonia 2

1 An equilibrium reaction is one which can go in both directions. The reactions in each direction happen at the same rate. Equilibrium can only be achieved in a closed system. The reaction below is an example of an equilibrium reaction.
   $N_2(g) + 3H_2(g) \rightleftharpoons 2NH_3(g)$
2 When an equilibrium reaction is subjected to a change in conditions, the position of the equilibrium reaction shifts so as to counteract the effect of the change. So if the pressure is increased in a gaseous reaction, the reaction that produces fewer molecules is favoured, as shown by the equation for the reaction.

C3 12–13 Levelled questions: The production of ammonia

Working to Grade E

1

Raw material	Source
nitrogen	separated from the air
hydrogen	obtained from natural gas or other sources

2

temperature (°C)	450
pressure (atmospheres)	200
catalyst	iron

Working to Grade C

3 1 – A, 2 – B, 3 – C, 4 – E, 5 – D
4 a True
  b In the reaction vessel, ammonia molecules break down to make nitrogen and hydrogen.
  c True
  d One mole of nitrogen reacts with three moles of hydrogen to make two moles of ammonia.
  e True

Working to Grade A*

5

Change	Effect on position of equilibrium		
	Shifts left	No change	Shifts right
increasing temperature			✓
decreasing pressure			✓
adding a catalyst		✓	

6 Increasing temperature shifts the equilibrium towards the endothermic reaction which absorbs more energy so tending to counteract the effect of increasing the temperature.

Decreasing pressure shifts the equilibrium towards the reaction that produces the higher number of molecules as shown by the symbol equation for the reaction.

There is no effect on the position of the equilibrium when a catalyst is added – catalysts change the activation energy of a reaction, and therefore its rate. They do not affect the position of the equilibrium.

7 a True
   b True
   c At equilibrium, the amounts of the products in the mixture remain the same.
   d At equilibrium, the amounts of the reactants in the mixture remain the same.
   e True
   f True
8 a The higher the pressure, the higher the yield of ammonia.
   b At this pressure, the yield is acceptable. At higher pressures, the yield would be greater, but the costs of building and maintaining the plant would be much more.

C3 12–13 Examination questions: The production of ammonia

1 a i   The symbol shows that the reaction is reversible. (1)
      ii 3, 2 (1)
   b i   So that they are not wasted (1) and to reduce costs. (1)
      ii To increase the rate of the reaction. (1)
      iii 1 mark awarded for each reason, up to a maximum of 2.
          • To reduce the energy requirements of the process – if energy was not transferred from the cooling gases, it would have to be supplied to the reaction vessel from another source.
          • To reduce the temperature of the gases – ammonia condenses at a higher temperature than hydrogen and nitrogen, and so can be separated from the mixture.
   c i   As temperature increases, the yield of ammonia decreases. (1)
      ii A temperature of 450 °C is chosen for the Haber process. At this temperature, the percentage yield is smaller than it is at lower temperatures. But the rate of the reaction increases as temperature increases. (1) At 450 °C the yield and rate are both acceptable. (1) The environmental impact and energy requirements would be less at lower temperatures, but the rate would be too slow. (1)
2 a 1 mark for each of the two true statements identified. True statements: In the equilibrium mixture, sulfur dioxide and oxygen are reacting to make sulfur trioxide.
         In the equilibrium mixture, sulfur trioxide is decomposing to make sulfur dioxide and oxygen.

2 b **Here is a guide to the marking for an essay-style question:**
      To gain 5/6 marks – All information in the answer should be relevant, clear, organised, and presented in a structured and coherent format. Specialist terms will be used appropriately. Few, if any, errors in grammar, punctuation, and spelling. Answer should include 5 or 6 points from those below.
      To gain 3/4 marks – Most of the information should be relevant and presented in a structured and coherent format. Specialist terms will usually be used correctly. There may be occasional errors in grammar, punctuation, and spelling. Answer should include 3 or 4 points of those listed below.
      To gain 1/2 marks – Answer will be simplistic. There may be limited use of specialist terms. Errors of grammar, punctuation, and spelling will prevent communication of the science. Answer should include 1 or 2 points of those listed below.
      Points to include:
      • If a reaction at equilibrium is subjected to a change, the position of the equilibrium changes so as to tend to counteract the change.
      • If the pressure increases, the equilibrium shifts towards the right.
      • This is because there are fewer molecules shown on the right of the equation.
      • If the pressure decreases, the equilibrium shifts towards the left.
      • If the temperature increases, the equilibrium shifts towards the endothermic reaction (left).
      • This is because the endothermic reaction absorbs the extra energy, so tending to counteract the effect of increasing temperature.
      • If the temperature decreases, the equilibrium shifts towards the exothermic reaction (right).
      • This is because the exothermic reaction releases energy, so tending to counteract the effect of decreasing the temperature.

C3 14: Alcohols

1 –O–H
2

Reacts with...	Products
oxygen in combustion reactions	carbon dioxide and water
sodium	hydrogen and sodium ethoxide
oxygen from oxidising agents or by action of microbes	ethanoic acid

3 $2CH_3OH + 3O_2 \rightarrow 2CO_2 + 4H_2O$

C3 15: Carboxylic acids

1

React with...	Products
oxygen in combustion reactions	carbon dioxide and water
carbonates	salt and carbon dioxide and water
alcohols	ester and water

2 To make vinegar; as food additives; as a component of vitamin tablets (ascorbic acid); as a painkiller (aspirin).

3 The pH will be higher for ethanoic acid. pH is a measure of hydrogen ion concentration. The greater the hydrogen ion concentration, the lower the pH. The hydrogen ion concentration is less in ethanoic acid because ethanoic acid does not ionise completely in water. It is a weak acid. Hydrochloric acid does ionise completely in water. It is a strong acid.

C3 16: Esters

1

ethyl ethanoate

2 ethanol + ethanoic acid → ethyl ethanoate + water

3 Esters are used as solvents in cosmetics and food flavourings.

C3 14–16 Levelled questions: Alcohols, carboxylic acids, and esters

Working to Grade E

1

Compound	Use
ethanol	alcoholic drinks
ethanoic acid	vinegar
pentyl pentanoate	flavouring

2 neutral, hydrogen, acidic, carbon dioxide.

3 Any correct answers are acceptable as examples, including those in the table below.

Name of group of organic compounds	Examples
alcohols	ethanol methanol
carboxylic acids	ethanoic acid propanoic acid
esters	ethyl ethanoate ethyl propanoate

4 functional, propanol, homologous fuels, solvents, drinks, ethanol, ethanoic, oxidising, microbes

Working to Grade C

5 **a** Carboxylic acids have the functional group –COOH

b Carboxylic acids react with alcohols to produce esters.

c The molecular formula of methanoic acid is HCOOH.

d True

e True

f Propyl propanoate is an ester.

6

Name	Molecular formula	Structural formula
methanol	CH₃OH	
ethanol	CH₃CH₂OH	
propanol	CH₃CH₂CH₂OH	
methanoic acid	HCOOH	
ethanoic acid	CH₃COOH	
propanoic acid	CH₃CH₂COOH	

7 **a** ethanol, propanoic acid

b propanol, ethanoic acid

c methanoic acid, methanol

d ethanoic acid, ethanol

8 **a** ethanol + oxygen → carbon dioxide + water

b methanol + sodium → sodium methoxide + hydrogen

c ethanol + sodium → sodium ethoxide + hydrogen

d propanol + oxygen → carbon dioxide + water

e ethanoic acid + sodium carbonate → sodium ethanoate + carbon dioxide + water

f propanoic acid + calcium carbonate → calcium propanoate + carbon dioxide + water

g ethanol + ethanoic acid → ethyl ethanoate + water

h propanol + propanoic acid → propyl propanoate + water

Working to Grade A*

9 **a** $2CH_3OH + 3O_2 \rightarrow 2CO_2 + 4H_2O$

b $CH_3CH_2OH + 3O_2 \rightarrow 2CO_2 + 3H_2O$

10 Ethanoic acid does not ionise completely in water. This means it is a weak acid.

11 a Acid A – it has the lowest pH, meaning it is the strongest acid of those in the table.
   b Acid C is a weaker acid than acid B, because the pH of acid C is higher.

C3 14–16 Examination questions: Alcohols, carboxylic acids, and esters

1 a

ethyl ethanoate (1)

   b alcohols and carboxylic acids. (1)
   c To act as a catalyst *or* to speed up the reaction. (1)
   d It has an interesting flavour. (1)
2 Uses: fuels, solvents, alcoholic drinks.
   Here is a guide to the marking for an essay-style question:
   To gain 5/6 marks – All information in the answer should be relevant, clear, organised, and presented in a structured and coherent format. Specialist terms will be used appropriately. Few, if any, errors in grammar, punctuation, and spelling. Answer should include 5 or 6 points from those below.
   To gain 3/4 marks – Most of the information should be relevant and presented in a structured and coherent format. Specialist terms will usually be used correctly. There may be occasional errors in grammar, punctuation, and spelling. Answer should include 3 or 4 points of those listed below.
   To gain 1/2 marks – Answer will be simplistic. There may be limited use of specialist terms. Errors of grammar, punctuation, and spelling will prevent communication of the science. Answer should include 1 or 2 points of those listed below.
   Points to include (for alcoholic drinks):
   • Social disadvantage of drinking alcohol – it slows down reaction times, increasing risks of road accidents.
   • Social disadvantage of drinking alcohol – it makes people forgetful, confused, and more likely to act foolishly.
   • Social disadvantage of drinking alcohol – it may cause vomiting, unconsciousness, or death.
   • Social advantage of drinking alcohol – it makes people feel relaxed for a short time.
   • Economic advantages of drinking alcohol – profitable for drinks companies.
   • Economic advantage of drinking alcohol – taxes provide income for government.

• Economic disadvantage of drinking alcohol – treating alcohol-related health problems is costly.
• Economic disadvantage of drinking alcohol – dealing with alcohol-related crime is costly.
• Statement evaluating the benefits and costs of drinking alcohol.
Note: There are many other possible correct answers to this question.

3 a

(1)

   b carbon dioxide (1)
   c Compound A (1)
   d i Only some of its molecules are ionised in solution. (2)
      ii $0.1\,mol/dm^3$ ethanoic acid, (1) since this is a carboxylic acid. In a solution of a carboxylic acid, only some of the ethanoic acid molecules are ionised. (1)

How Science Works: Analysis, ammonia, and organic chemistry

1 Questions could include: How many repeats should I do for each acid concentration? What range of acid concentration should I use? What intervals between the acid concentrations would be best?
2 Independent variable – acid concentration. Dependent variable – volume of acid required to neutralise $25\,cm^3$ of $1.0\,mol/dm^3$ sodium hydroxide solution.
3 The measurements must measure only the volumes of acid required, and should be as precise as possible.
4 Suzanna must control all the variables that might affect the dependent variable, including the concentration of sodium hydroxide solution, and the volume of sodium hydroxide solution, and the indicator chosen.
5 At acid concentration 1.0, the data is clustered closely. At acid concentration 1.5, the data is clustered less closely. This indicates that the data for concentration 1.0 is more precise.
6 Suzanna might have made mistakes in the investigation; she might have read some values incorrectly; she might have found it difficult to judge the exact level of the acid meniscus on the burette scale; the equipment might not have been clean.

Appendices

Periodic table

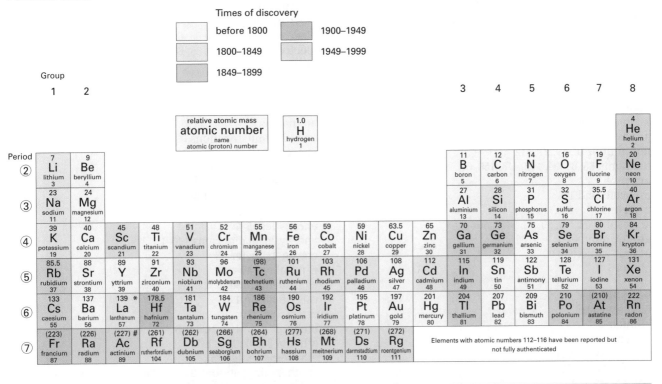

Times of discovery

☐ before 1800	☐ 1900–1949
☐ 1800–1849	☐ 1949–1999
☐ 1849–1899	

Group

| | | | | | | |
| 1 | 2 | | 3 | 4 | 5 | 6 | 7 | 8 |

| relative atomic mass |
| **atomic number** |
| name |
| atomic (proton) number |

| 1.0 |
| **H** |
| hydrogen |
| 1 |

	4
	He
	helium
	2

Period																			
2	7 **Li** lithium 3	9 **Be** beryllium 4											11 **B** boron 5	12 **C** carbon 6	14 **N** nitrogen 7	16 **O** oxygen 8	19 **F** fluorine 9	20 **Ne** neon 10
3	23 **Na** sodium 11	24 **Mg** magnesium 12											27 **Al** aluminium 13	28 **Si** silicon 14	31 **P** phosphorus 15	32 **S** sulfur 16	35.5 **Cl** chlorine 17	40 **Ar** argon 18
4	39 **K** potassium 19	40 **Ca** calcium 20	45 **Sc** scandium 21	48 **Ti** titanium 22	51 **V** vanadium 23	52 **Cr** chromium 24	55 **Mn** manganese 25	56 **Fe** iron 26	59 **Co** cobalt 27	59 **Ni** nickel 28	63.5 **Cu** copper 29	65 **Zn** zinc 30	70 **Ga** gallium 31	73 **Ge** germanium 32	75 **As** arsenic 33	79 **Se** selenium 34	80 **Br** bromine 35	84 **Kr** krypton 36
5	85.5 **Rb** rubidium 37	88 **Sr** strontium 38	89 **Y** yttrium 39	91 **Zr** zirconium 40	93 **Nb** niobium 41	96 **Mo** molybdenum 42	(98) **Tc** technetium 43	101 **Ru** ruthenium 44	103 **Rh** rhodium 45	106 **Pd** palladium 46	108 **Ag** silver 47	112 **Cd** cadmium 48	115 **In** indium 49	119 **Sn** tin 50	122 **Sb** antimony 51	128 **Te** tellurium 52	127 **I** iodine 53	131 **Xe** xenon 54
6	133 **Cs** caesium 55	137 **Ba** barium 56	139 **La** * lanthanum 57	178.5 **Hf** hafnium 72	181 **Ta** tantalum 73	184 **W** tungsten 74	186 **Re** rhenium 75	190 **Os** osmium 76	192 **Ir** iridium 77	195 **Pt** platinum 78	197 **Au** gold 79	201 **Hg** mercury 80	204 **Tl** thallium 81	207 **Pb** lead 82	209 **Bi** bismuth 83	210 **Po** polonium 84	(210) **At** astatine 85	222 **Rn** radon 86
7	(223) **Fr** francium 87	(226) **Ra** radium 88	(227) **Ac** # actinium 89	(261) **Rf** rutherfordium 104	(262) **Db** dubnium 105	(266) **Sg** seaborgium 106	(264) **Bh** bohrium 107	(277) **Hs** hassium 108	(268) **Mt** meitnerium 109	(271) **Ds** darmstadtium 110	(272) **Rg** roentgenium 111	Elements with atomic numbers 112–116 have been reported but not fully authenticated						

*58–71 Lanthanides

| 140 **Ce** cerium 58 | 141 **Pr** praseodymium 59 | 144 **Nd** neodymium 60 | (145) **Pm** promethium 61 | 150 **Sm** samarium 62 | 152 **Eu** europium 63 | 157 **Gd** gadolinium 64 | 159 **Tb** terbium 65 | 163 **Dy** dysprosium 66 | 165 **Ho** holmium 67 | 167 **Er** erbium 68 | 169 **Tm** thulium 69 | 173 **Yb** ytterbium 70 | 175 **Lu** lutetium 71 |

#90–103 Actinides

| 232 **Th** thorium 90 | 231 **Pa** protactinium 91 | 238 **U** uranium 92 | 237 **Np** neptunium 93 | 239 **Pu** plutonium 94 | 243 **Am** americium 95 | 247 **Cm** curium 96 | 247 **Bk** berkelium 97 | 252 **Cf** californium 98 | (252) **Es** einsteinium 99 | (257) **Fm** fermium 100 | (258) **Md** mendelevium 101 | (259) **No** nobelium 102 | (260) **Lr** lawrencium 103 |

Reactivity series of metals

Potassium	most reactive ↑
Sodium	
Calcium	
Magnesium	
Aluminium	
Carbon	
Zinc	
Iron	
Tin	
Lead	
Hydrogen	
Copper	
Silver	
Gold	
Platinum	least reactive ↓

(elements in italics, though non-metals, have been included for comparison)

Formula of some common ions

Name	Formula	Name	Formula
Hydrogen	H^+	Chloride	Cl^-
Sodium	Na^+	Bromide	Br^-
Silver	Ag^+	Fluoride	F^-
Potassium	K^+	Iodide	I^-
Lithium	Li^+	Hydroxide	OH^-
Ammonium	NH_4^+	Nitrate	NO_3^-
Barium	Ba^{2+}	Oxide	O^{2-}
Calcium	Ca^{2+}	Sulfide	S^{2-}
Copper(II)	Cu^{2+}	Sulfate	SO_4^{2-}
Magnesium	Mg^{2+}	Carbonate	CO_3^{2-}
Zinc	Zn^{2+}		
Lead	Pb^{2+}		
Iron(II)	Fe^{2+}		
Iron(III)	Fe^{3+}		
Aluminium	Al^{3+}		

Index

UNIVERSITY PRESS

Great Clarendon Street, Oxford OX2 6DP

Oxford University Press is a department of the University of Oxford.
It furthers the University's objective of excellence in research,
scholarship, and education by publishing worldwide in

Oxford   New York

Auckland Cape Town Dar es Salaam Hong Kong Karachi
Kuala Lumpur Madrid Melbourne Mexico City Nairobi
New Delhi Shanghai Taipei Toronto

With offices in
Argentina Austria Brazil Chile Czech Republic France Greece
Guatemala Hungary Italy Japan Poland Portugal Singapore
South Korea Switzerland Thailand Turkey Ukraine Vietnam

Oxford is a registered trade mark of Oxford University Press
in the UK and in certain other countries.

British Library Cataloguing in Publication Data

Data available

ISBN 978-0-19-913604-9

10 9 8 7 6 5 4 3 2

Printed in Great Britain by Bell and Bain Ltd. Glasgow

Paper used in the production of this book is a natural, recyclable product
made from wood grown in sustainable forests. The manufacturing process
conforms to the environmental regulations of the country of origin.